D1233072

RESEARCH AND PERSPECTIVES IN NEUROSCIENCES

Fondation Ipsen

Editor

Yves Christen, Fondation Ipsen, Paris (France).

Springer

Berlin
Heidelberg
New York
Barcelona
Hong Kong
London
Milan
Paris
Singapore
Tokyo

J. Grafman Y. Christen (Eds.)

Neuronal Plasticity: Building a Bridge from the Laboratory to the Clinic

With 44 Figures and 3 Tables

 Springer

Grafman, J., Ph.D.
Cognitive Neuroscience Section
Medical Neurology Branch
National Institute of Neurological Disorders and Stroke
National Institutes of Health
Building 10, Room 5C205
10 Center Drive MSC 1440
Bethesda, MD 20892–1440
USA

Christen, Y., Ph.D.
Fondation IPSEN
Pour la Recherche Thérapeutique
24, rue Erlanger
75781 Paris Cedex 16
France

ISSN 0945–6082
ISBN 3–540-64357–5   Springer-Verlag Berlin Heidelberg New York

Library of Congress Cataloging-in-Publication Data
Neuronal plasticity : building a bridge from the laboratory of the clinic / J. Grafman, Y. Christen (eds.). p. cm. – (Research and perspectives in neurosciences, ISSN 0945-6082) Includes bibliographical references and index. ISBN 3-540-64357-5 (alk. paper) 1. Neuroplasticity–Congresses. 2. Brain damage–Pathophysiology-Congresses. 3. Brain damage–Patients–Rehabilitation–Congresses. I. Grafman, Jordan. II. Christen, Yves. III. Series. QP363.3.N4485 1998 616.8'047–dc21 98-38384 CIP

Production: PRO EDIT GmbH, D-69126 Heidelberg
Cover design: design & production, D-69121 Heidelberg
Typesetting: Mitterweger Werksatz GmbH, Plankstadt
SPIN: 10551697   27/3136 – 5 4 3 2 1 0 – Printed on acid-free paper

Preface

The brain is an instrument of change. When we learn and when we recover from brain damage, the brain acts as a dynamic organ adapting itself to our interests or our efforts to regain essential abilities. While we are far from a complete understanding of the neuroplastic operations of brain networks, the last two decades have seen an explosion of knowledge in this area of research. Despite the excitement surrounding such work, much of the research has not been translated into practical interventions that could be introduced into the school, work place or rehabilitation clinic. The goal of the meeting of the Fondation Ipsen (Paris, October 6, 1997) which spawned this volume was to place recent basic research on neuroplasticity at the doorstep of human clinical research. We believe that we accomplished that goal and this volume provides the evidence for that belief.

The chapters in this volume reflect the range of work presented at the meeting. The presentations ranged from addressing the plasticity of nicotinic receptors to connectionist modeling of relearning in dyslexics. While we planned the meeting to emphasize clinical applications of neuroplasticity research, we felt it was first necessary to lay a foundation of basic science. In this regard, the chapters by Jean-Pierre Changeux and Bryan Kolb indicate the progress that is being made in determining the neural underpinnings of plasticity and the effects that experience can have on neuroplasticity. Clinical research examining neuroplasticity is both varied and exciting. Ronald Melzack reports on current thinking on the development of chronic pain following injury and how neuroplastic changes may contribute to this phenomenon. Josef Rauschecker, Salvatore Aglioti and Alvaro Pascual-Leone discuss auditory, somatensory, and visual system plasticity respectively while Leonardo Cohen indicates the key factors that contribute to the functional relevance of neuroplastic change. Although neuroplasticity is typically demonstrated by changes in overt behavior, Jean Decety points out that changes in mental imagery ability can also provide evidence of neuroplastic change. Richard Frackowiak's plethora of functional neuroimaging studies demonstrating changes in patterns of brain activity during learning and recovery of function are powerful witnesses to brain plasticity in action. David Plaut's application of artificial intelligence techniques to model learning and relearning of behavior introduces an important new tool for both the modeling and prediction of brain and behavioral response to injury. Although he stresses connectionist approaches in this chapter, other AI modeling techniques may also prove useful in this regard. In his chapter, Jordan Grafman proposes four major forms that cognitive neuroplasticity may take. Given the exponentially increasing number of studies investigating neuroplasticity, searching for the basic organizing principles of cognitive neuroplasticity is a necessary exercise. Finally, Michael Merzenich indicates how case-based examples of neuroplasticity research has been able to span the bridge between basic research and the rehabilitation clinic and dramatically improve the reading performance of children with certain forms of dyslexia.

Besides the chapters contained in this volume, Antonio Damasio, Hannah Damasio, and François Chollet as session chairs kept the conversational laminar flow of the meeting moving along at a timely pace while offering their own incisive comments and questions on the presentations that no doubt helped the authors improve the quality of their chapters. The poster sessions were also quite interesting and showed that clinical research activity in France and Europe are at the cutting edge of neuroplasticity research. In addition, the targeted and enthusiastic questions of many of the clinicians in the audience demonstrated that they are eager to put into practice the experimental procedures advocated by the meeting participants.

Jacqueline Mervaillie and the Fondation Ipsen were, as usual, the most gracious, creative and organized of hosts. If there was a Copa Fondation Mundial, both from a logistical and cultural point of view, the Fondation Ipsen would be taking home the world cup. It is always a privilege to work with them.

One of us (JG) first had the opportunity to observe and study recovery of function and brain neuroplasticity during his graduate training at the University of Wisconsin-Madison. His mentor during that time, Charles G. Matthews, greatly encouraged him and gave him the intellectual tools to carry on his work beyond graduate school. Dr Matthews unexpectedly died recently. For his intellectual and moral contribution to neuropsychology and his personal commitment to mentoring, we honor his memory with this volume.

To Irene and Phyllis,   Jordan Grafman

Acknowledgements: The editors also wish to express their gratitude to Mary Lynn Gage for her editorial assistance.

Contents

Some Neurological Principles Relevant to the Origins of –
and the Cortical Plasticity-Based Remediation of –
Developmental Language Impairments

List of Contributors

Aglioti, S.
Dipartimento di Scienze Neurologiche e della Visione, Sezione di Fisiologia Umana, Strada le Grazie 8, Universitá di Verona, 37134 Verona, Italy

Catalá, M.D.
Unidad de Neurobiologia, Dept. Fisiologia, Universidad de Valencia and Instituto Cajal, Consejo Superior de Investigaciones Científicas, Avenida Blasco Ibañez 17, 46010 Valencia, Spain

Celnik, P.
Human Cortical Physiology Unit, National Institute of Neurological Disorders and Stroke, National Institutes of Health, Building 10, Room 5N234–1430, Bethesda, MD 20892, USA

Changeux, J.-P.
Institut Pasteur, 25 rue du Docteur Roux, 75015 Paris, France

Chen, R.
Human Cortical Physiology Unit, National Institute of Neurological Disorders and Stroke, National Institutes of Health, Building 10, Room 5N234–1430, Bethesda, MD 20892, USA

Coderre, T.J.
Institut de Recherche Clinique de Montreal, Montreal, Quebec, Canada LCH2W IR7

Cohen, L.G.
Human Cortical Physiology Unit, National Institute of Neurological Disorders and Stroke, National Institutes of Health, Building 10, Room 5N234–1430, Bethesda, MD 20892, USA

Decety, J.
Mental Processes and Brain Activation – Inserm Unit 280, 151 cours Albert Thomas, 69424 Lyon Cedex 03 and Cermep, 59 Bld. Pinel, 69003 Lyon, France

Frackowiak, R.S.J.
Wellcome Department of Cognitive Neurology, Institute of Neurology, 12 Queen Square, London WCIN 3BG, UK

Grafman, J.
Cognitive Neuroscience Section, NIH/NINDS/MNB, Building 10, Room 5C205, 10 Center Drive MSC 1440 Bethesda, MD 20892–1440, USA

Hamilton, R.
Laboratory for Magnetic Brain Stimulation, Department of Neurology, Beth Israel Deaconess Medical Center, 330 Brookline Ave, KS 452, Boston, MA 02215, USA

Jenkins, W.M.
Scientific Learning Corporation, 1995 University Avenue, Berkeley, CA 94704–1074
and Keck Center for Integrative Neurosciences, University of California at San Francisco, San Francisco, CA 94143–0732, USA

Katz, J.
Department of Psychology, Toronto Hospital, Toronto, Ontario, Canada

Keenan, J.
Laboratory for Magnetic Brain Stimulation, Department of Neurology, Beth Israel Deaconess Medical Center, 330 Brookline Ave, KS 452, Boston, MA 02215, USA

Kolb, B.
Department of Psychology and Neuroscience, University of Lethbridge, Lethbridge, AB, Canada, T1K 3M4

Léna, C.
Laboratoire de Neurobiologie Moléculaire, Institut Pasteur, 25–28 rue du Dr. Roux, 75724 Paris, Cedex 15, France

Litvan, I.
Neuropharmacology Unit, Defense and Veterans Head Injury Program, Henry M. Jackson Foundation, Rockville, MD, USA

Melzack, R.
Institut de Recherche Clinique de Montreal, Montreal, PQHSA 1B1 Quebec, Canada

Merzenich, M.M.
Keck Center for Integrative Neurosciences, University of California at San Francisco, San Francisco, CA 94143–0732, USA and Scientific Learning Corporation, 1995 University Avenue, Berkeley, CA 94704–1074, USA

Miller, S.
Scientific Learning Corporation, 1995 University Avenue, Berkeley, CA 94704–1074, USA

Pascual-Leone, A.
Laboratory for Magnetic Brain Stimulation, Department of Neurology, Beth Israel Deaconess Medical Center, 330 Brookline Ave, KS 452, Boston, MA 02215, USA and Unidad de Neurobiologia, Dept. Fisiologia, Universidad de Valencia and Instituto Cajal, Consejo Superior de Investigaciones Científicas, Avenida Blasco Ibañez 17, 46010 Valencia, Spain

Peterson, B.
Scientific Learning Corporation, 1995 University Avenue, Berkeley, CA 94704–1074, USA

Plaut, D.C.
Departments of Psychology and Computer Science, Center for the Neural Basis of Cognition, Carnegie Mellon University, Mellon Institute 115-CNBC, 4400 Forbes Avenue, Pittsburgh, PA 15213–2683, USA

Rauschecker, J.P.
Georgetown Institute for Cognitive and Computational Sciences, Georgetown University Medical Center, Washington, DC 20007, USA

Tallal, P.
Center for Molecular and Behavioral Neuroscience, Rutgers University, 197 University Avenue, Newark, NJ 07102 and Scientific Learning Corporation, 1995 University Avenue, Berkeley, CA 94704–1074

Tormos, J.M.
Unidad de Neurobiologia, Dept. Fisiologia, Universidad de Valencia and Instituto Cajal, Consejo Superior de Investigaciones Científicas, Avenida Blasco Ibañez 17, 46010 Valencia, Spain

Vaccarino, A.L.
Department of Psychology, University of New Orleans, New Orleans, Louisiana 70148, USA

Pathological Mutations of Nicotinic Receptors and Nicotine-Based Therapies for Brain Disorders

C. Léna* and J.-P. Changeux*

Summary

Nicotinic acetylcholine receptors are allosteric ligand-gated ion channels present in muscle and brain. Recent studies suggest that mutations altering their functional properties may produce congenital myasthenia and familial frontal lobe epilepsy. Current research also indicates that although nicotinic ligands often possess addictive properties, they could serve as therapeutic agents for Alzheimer's disease and Tourette's syndrome, as well as for schizophrenia.

Introduction

Nicotinic acetylcholine receptors (nAChRs) were the first neurotransmitter receptors to be identified biochemically and functionally, in part because it was found that high amounts of the protein are present in *Torpedo* electric organ (reviewed in Changeux 1980). Recombinant DNA technologies permitted the demonstration that the structal and functional properties of this allosteric membrane protein, to a large extent, parallel those of brain nAChRs, thus opening the field to research on human brain pathologies.

In this review, two aspects of nAChRs relevant to medicine are presented. First, the possibility that point mutations in muscle and brain nAChRs may produce congenital myasthenia and familial epilepsies. The phenotype of mutated nAChRs is interpreted in terms of changes in the properties of the allosteric transitions. Second, nicotinic drugs, despite their addictive properties, could potentially alleviate neurological and psychiatric disorders.

Nicotinic Receptors as Allosteric Membrane proteins

The nAChRs are a family of ligand-gated ion channels that are differentially expressed in skeletal muscle and nerve cells (reviewed in Galzi and Changeux 1995; Role and Berg 1996). They form 300 kDa transmembrane hetero- (or homo-) pentamers from a repertoire of 16 known different types of subunits

* Laboratoire de Neurobiologie Moléculaire, Institut Pasteur, 25–28 rue du Dr Roux, 75724 Paris, Cedex 15, France
Reprinted from current opinion in Neurobiology, 1997, 7, 674–682. With permission

J. Grafman / Y. Christen (Eds.)
Neuronal Plasticity:
Building a Bridge from the Laboratory to the Clinic
© Springer-Verlag Berlin Heidelberg New York 1999

referred as α1–α9, β1–β4, γ, δ, and ε. The subunits are regularly distributed around an axis of quasi-symmetry delineating the ion channel (Figure 1 a). Each subunit contains a large amino-terminal hydrophilic domain exposed to the synaptic cleft, followed by three transmembrane segments (MI–M3), a large intracellular loop, and a carboxy-terminal transmembrane segment (M4). Acetylcholine-binding sites are located at the interface between α and non-α subunits in the amino-terminal regions (Galzi and Changeux 1995; Tsigelny et al. 1997). They include a principal component of three loops A, B and C and a complementary component of at least two loops D and E on the non-α subunit; in homo-oligomeric receptors, the two components are carried by identical subunits. A wide diversity of binding properties results from the combinatorial diversity of the active site structure. The ion channel is lined by the M2 segment from each of the five subunits. Neuronal nAChRs are more permeable to calcium ions than muscle nAChRs (neuronal nAChRs: pCa/pNa values from 15 to 0.5, depending on the subunit composition; muscle nAChRs: pCA/pNA values of about 0.2) (Galzi and Changeux 1995; Role and Berg 1996).

Muscle nAChRs have a fixed composition $[α1]_2[β1]$ [δ] or [γ or ε] in vertebrates. Neuronal nAChRs are composed of neuron-specific subunits homologous to the muscle subunits. To date, ten neuronal subunits have been identified in mammals (α2–α7, α9, β2–β4). Of the more than 20,000 possible combinations of subunits, only a few yield functional receptors. The α7 and α9 subunits from functional homo-oligomers when expressed in *Xenopus* oocytes, whereas the α2–α4 subunits produce hetero-oligomers with β2 or the β4 subunit (Figure 1 c) (reviewed in McGehee and Role 1995)). The rat α6 subunit assembles into functional hetero-oligomers with the human β4 subunit (Gerzanich et al. 1997), but evidence is still lacking for a functional nAChR containing the human α6 subunit. The α5 subunit can associate with α3β2/4 and α4β2subunits and thus from hetero-oligomers with three different subunits (Ramirez Latorre et al. 1996; Wang et al. 1996) (Figure 1 c). The contribution of the β3 subunit to a functional nAChR still awaits demonstration, but its sequence homology with α5 suggests that it possesses a similar function (Le Novère et al. 1996).

Upon application of nicotinic agonists, both muscle and neuronal nAChRs undergo fast activation leading to a open-channel state, as well as a slow desensitization reaction leading to a closed-channel state refractory to activation. Activation and desensitization of muscle and brain nAChRs correspond to transitions between a small number of discrete structural states with distinct binding properties and ion channel conductance (Changeux 1990).

Consistent with the allosteric two-state model and its extension to membrane receptors (Changeux 1990, Edelstein et al. 1996), the different conformational states of nAChRs may exist in the absence of nicotinic ligands, and allosteric effectors cooperatively modify the equilibrium and kinetic constants for the transitions between the states (Figure 1 b). The pharmacological and kinetic characteristics of these states depend upon the subunit composition of the receptor. Indeed, the two main subtypes of brain nAChRs differ strikingly: the human α4β2 hetero-oligomer and α7 homo-oligomer receptors have, respectively, a low

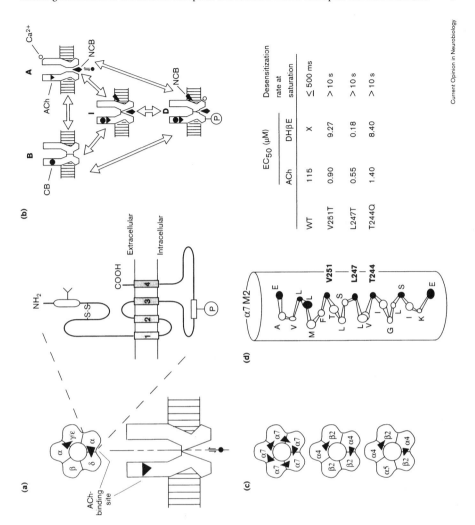

Fig. 1. Nicotinic acetylcholine receptors (nAChRs) are ligand-gated ion channels with allosteric properties (Changeux 1990). **(a)** Muscle and *Torpedo* nAChRs are pentameric oligomers. The five homologous subunits are organized around an axis of quasi-symmetry perpendicular to plane of the plasma membrane that delineates the ion channel pore. Each subunit exhibits a similar transmembrane organization sketched on the right. The binding sites are located at the interface of the extracellular amino-terminal domains of the subunits, and the ion channel is lined by the M2 (indicated by a '2' on the figure) transmembrane segment. ACh, acetylcholine; P, phosphorylation site. **(b)** The nAChRs undergo allosteric transitions between a small number of states; resting (B), active (A) and desensitized (1 and D) (Changeux 1990; Edelstein et al. 1996). Various ligands preferentially bind to different states, as indicated; CB, competitive blockers; NCB, noncompetitive blockers. **(c)** Putative organization of three different types of neuronal nAChRs: homopentamers of α7 subunits, heteropentamers of α4 and β2, and heteropentamers of α4, β2 and α5. **(d)** The M2 transmembrane segment is putatively organized in an α-helix. Mutations of residues in M2 facing the channel pore may increase the apparent affinity for acetylcholine, convert the antagonist dihydro-β-erythroidine (DHPE) into an agonist and drastically slow the desensitization rate of mutated α7-nAChRs. WT, wild-type. Adapted from (Devillers-Thiery et al. 1992).

and a high EC_{50} for nicotine (0.3-5 µM versus 40-110 µM); at saturation, they desensitize, respectively, in the 10s and in the 10-100 ms range (or below) (Peng et al. 1994; Gerzanich et al. 1995; Gopalakrishnan et al. 1995; Buisson et al. 1996; Chavez Noriega et al. 1997). The kinetic constants governing the ligand binding and the transitions between the different states (14 independent rate constants for a four state model) have been estimated for muscle nAChRs (Edelstein et al. 1996), and the analysis has been extended to neuronal nAChRs mutants (Edelstein and Changeux 1996).

Site-directed mutagenesis of affinity-labelled residues in the channel and active site domains revealed that mutations of single amino acids can modify multiple functions of the nAChR (Revah et al. 1991; Galzi et al. 1991; Devillers-Thiery et al. 1992; Bertrand et al. 1992; Labarca et al. 1995). For instance, mutations in the M2 channel-lining region – α7-Thr244→Gln (α7T244Q), α7-Leu247→Thr (α7L247T), α7-Val251→Thr (α7V251T) produce a 100-fold increase in apparent affinity for agonists, a loss of desensitization and a conversion of competitive antagonists to agonists (Figure 1 d) (Revah et al. 1991; Devillers-Thiery et al. 1992; Bertrand et al. 1992; Labarca et al. 1995 reviewed in Galzi and Changeux 1995). The allosteric model accounts for these pleiotropic phenotypes. Different classes of phenotypes may be distinguished by selective changes in the binding properties (K phenotype), in the biological activity of the ion channel (γ phenotype), or in the isomerization constants between receptor conformations (L phenotype) (Galzi et al. 1996).

The neuronal nAChR subunits are expressed differentially in the brain. *In situ* hybridization in rat brain shows that α4, α7 and β2 are widely expressed, that α3 and α5 are less ubiquitous, and that α2, α6, β3 and β4 are only expressed in a few brain structures (Table 1). In contrast, α3 and β4 are the most abundant nAChR subunits in the autonomic peripheral nervous system (Zoli et al. 1995). As a consequence of such diversity in function and distribution, neuronal nAChRs contribute to a wide array of brain functions (Changeux et al. 1996). Conversely, dysfunction of different single nAChR subunits may produce diverse symptoms.

Congenital Myasthenia and Familial Epilepsies Result from nAChR Point Mutations

Genetic analysis of several human (and animal) pathologies has revealed nAChR mutations yielding pleiotropic phenotypes (Figure 2). The mutations are homologous or even identical to those initially studied in reconstituted α7 homo-oligomers, and their phenotype may be also interpreted in terms of the allosteric model.

Mutation in the *deg3* gene coding for a putative nAChR subunit of the nematode *Caenorhabditis elegans* results in neurodegeneration (Treinin and Chalfie 1995). This mutation, Ile293→Asn (I293N; single letter amino acid code for mutations will continue to be used below), probably causes a 'increase-of-function' similar to the vertebrate α7V251T mutation (Devillers-Thiery et al. 1992). The neurotoxicity of the mutation could plausibly arise from a large toxic influx of calcium associated with a nondesensitizing and/or spontaneously open nAChR channel (Revah et al. 1991; Devillers-Thiery et al. 1992).

Table 1 Differential distribution of nAChR subunit mRNAs in rat brain.*

	α2	α3	α4	α5	α6	α7	β2	β3	β4
Telencephalon									
Olfactory bulb	+	++	+	++	–	++	++	–	+
Isocortex									
Layer II–III	–	–	+	+	–	+	++	–	–
Layer IV	–	+	+	+	–	+	++	–	–
Layer V	–	–	++	+	–	++	++	–	–
Layer VI	–	–	++	++	–	++	++	–	–
Hippocampal formation	(+)	(+)	+	+	–	+++	++	–	–
Striatum	–	–	–	–	–	–	+	–	–
Septum	–	–	+	–	–	+	+	–	–
Hypothalamus	–	–	+	–	–	–	+	–	–
Supraoptic nucleus	–	–	+	–	–	+++–	+	–	–
Diencephalon									
Pineal gland	–	+++	–	+	–	–	+	–	+++
Habenula	–	+++	++	+	(+)	(+)	++	++	+++
Thalamus	–	+	+++	–	+	–	+++	+	–
Mesencephalon									
Dopaminergic nuclei	–	(+)	++	++	+++	–	++	+++	–
Mesencephalic V nucleus	–	–	+	–	+	–	++	+++	–
Interpeduncular nucleus	++	+	+	++	+	(+)	++	+	+
Rhombencephalon									
Vestibular nuclei	–	–	+	+	–	++	+	–	–
Cerebellum	–	+	–	+	–	–	+	+	+
Locus coeruleus	–	(+)	–	–	+++	–	++	+++	(+)
Motor nuclei	–	+	+	+	–	–	++	–	–
NTS	–	++	(+)	+	–	–	++	–	++
Area postrema	–	++	++	+	–	–	++	–	+

* Data from Le Novère et al. 1996 and references therein. NTS, nucleus of the tractus solitarius.

In humans, maysthenia gravis is a sporadic disease caused by an auto-immune reaction directed against muscle nAChRs. However, some congenital myasthenic syndromes are associated with point mutations in muccle α1, β1 or ε subunits. Mutations reducing channel opening transition (e.g. εP121L) or affecting nAChR assembly (e.g. εR147L) cause myasthenic symptoms only when combined with a null mutation of the other allele (Ohno et al. 1996; Ohno et al. 1997). Null mutations cause myasthenic symptoms only when expressed on both alleles (Engel et al. 1996a). In accordance, animal models with a knock-out of the ε subunit express obvious myasthenic symptoms only in a homozygous genotype (Witzemann et al. 1996). In these myasthenic patients and animal models, neuro-

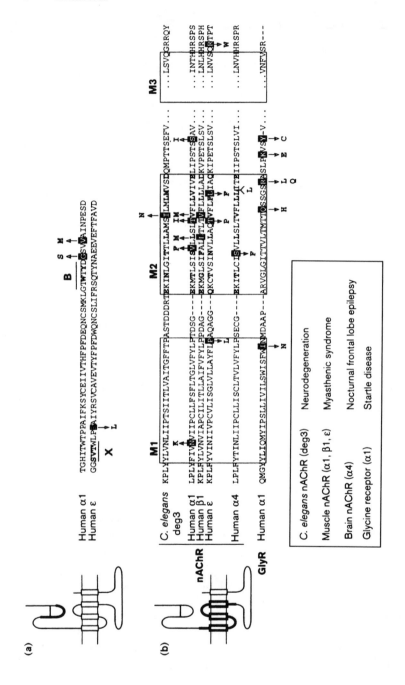

transmission is partially rescued at the neuromuscular junction by the persistence of expression of the fetal nAChR subunit γ.

Mutations increasing the time spent by nAChRs in the open state also produce myasthenic syndromes, even as heterozygous mutations (Ohno et al. 1995; Sine et al. 1995; Engels et al. 1996 b; Gomez et al. 1996; Croxen et al. 1997; Milone et al. 1997), but see (Ohno et al. 1997). Mutations causing such increase-of-function phenotypes occur near the ligand-binding region (α1G153S, α1V156M), in the transmembrane M2 segment delineating the ion channel (α1T254I, α1V249F, β1V266M, β1L262M, εL269F) or in adjacent regions (α1N217K, α1S269I, εP245L) (Figure 2). They may affect both intrinsic ligand-binding (K phenotype) and opening transition/desensitization (L phenotype) processes (see Galzi et al. 1996). Neighbouring mutations may produce different phenotypes; for instance, α1G153S slows agonist dissociation, whereas α1V156M decreases the rate of channel closing (Sine et al. 1995; Croxen et al. 1997). Some mutations (α1V249F, β1V266M, εT264P, εL269F) produce a high rate of spontaneous openings in the absence of ligand (Ohno et al. 1995; Engel et al. 1996 b; Milone et al. 1997), a phenotype consistent with the allosteric model on the basis of a shift of the allosteric equilibrium in favour of the open state (Edelstein et al. 1996). The confirmation that such increase-of-function mutations in muscle nAChRs are pathogenic was obtained recently in an animal model of transgenic mice carrying the εL269F mutation (Gomez et al. 1997).

Recently, some familial epilepsies have been linked to mutations in the α4 nAChR subunit. In an Australian family, an α4S248F mutation was found to produce autosomal dominant frontal lobe epilepsy (Steinlen et al. 1995). The mutated serine faces the channel pore, as initially demonstrated by chlorpromazine labelling in *Torpedo* nAChRs (see Galzi and Changeux 1995 and references therein). The mutation of the homologous residue in brain α7 nAChRs (α7T244Q) (Figure 1 d) causes drastic changes in the affinity for acetylcholine

Fig. 2. Pathogenic mutations affecting allosteric properties in the muscle and neuronal nAChRs and in the homologous glycine receptor (GlyR). Each disease is caused by a single mutation among the mutations indicated. The drawings on the left indicate (in bold) the protein domain concerned. The mutations are indicated by the single letter amino acid above/under the sequence with an arrow pointing from the wild-type highlighted residue. (a) Mutations near the ligand binding regions. B, B-loop in the principal component of the binding site; X, unnamed region in the complementary component of the binding site (Tsigelny et al. 1997). The mutations drawn are εP121L (Ohno et al. 1996)], α1 G153S (Sine et al. 1995; Croxen et al. 1997) and α1 V156M (Croxen et al. 1997). (b) Mutations in/near the transmembrane domains. Note the large number of mutations in the M2 region. The amino acids facing the channel pore are indicated in bold. The mutations in the nAChR subunit gene are *deg-3*-I293N (Treinin and Chalfie 1995), α1N217K (Engel et al. 1996 b), α1 V249F (Milone et al. 1997), α1T254I (Croxen et al. 1997), α1 S269l (Croxen et al. 1997), β1L262M (Gomez et al. 1996), β1V266M (Engel et al. 1996 b), εP245L (Ohno et al. 1997), εT264P (Ohno et al. 1995), εL269F (Engel et al. 1996 b; Gomez and Gammack 1995), εR311W (Ohno et al. 1997), α4S248F (Steinlein et al. 1995) and α4-776(ins3) (Steinlein et al. 1997). The mutations in the α1 glycine receptor gene are GlyRα1-I244N, GlyRα-Q266H, GlyRα1-R271Q, GlyRα1-R271L, GlyRα1-K276E and GlyRα1-K279C (references in Lynch et al. 1997). Alignments were performed with the Clustalw program of DG Higgins and PM Sharp.

and the desensitization properties of the nAChRs (Devillers-Thiery et al. 1992). In the human $\alpha 4$ gene, $\alpha 4S248F$ produces a twofold increase in the apparent affinity for acetylcholine and a fivefold increase in the desensitization rate of $\alpha 4\beta 2$ nAChRs (Weiland et al. 1996). In a Norwegian family, the same epileptic syndrome was linked to the insertion of a GCT triplet at nucleotide 776, resulting in the insertion of a leucine at codon 260 (Steinlen et al. 1997). In oocyte experiments, this insertion causes a 12-fold increase in the apparent affinity for acetylcholine of human $\alpha 4\beta 2$. As the insertion is adjacent to a pair of leucines previously identified as a critical element of calcium permeability (Bertrand et al. 1993), it might also reduce calcium permeability, though definitive evidence for this is lacking. It is unclear whether the phenotype of both these mutations, $\alpha 4S248F$ and $\alpha 4(776ins3)$, is attributable to a loss-of-function (increased desensitization rate and loss of calcium permeability) or an increase-of-function (increase in apparent affinity).

Point mutations that change allosteric properties occur in other ligand-gated ion channels, such as the glycine receptor $\alpha 1$ subunit. Human hereditary hyperekplexia is caused by mutations in the MI–M2 and M–M3 loops that lead to a dramatic reduction of efficacy of the agonist (see Lynch et al. 1997 and references therein). Overall, these results show that point mutations can cause either a loss-of-function or an apparent increase-of-function by altering the allosteric transitions of the nAChRs. Increase-of-function mutations occur frequently and may be as pathogenic as null mutations. As each allosteric state of the nAChR possesses a distinct pharmacological profile, one may anticipate the development of novel pharmacological agents targeted not only to a particular combination of subunits but to each of the diverse possible conformations of the various receptor oligomers.

Null Mutation of Neuronal nAChRs, Alzheimer's Disease and Memory

The role of defined nAChR subunits in brain frunction has been examined in knock-out animals. Mice lacking the most widely expressed $\beta 2$ subunit survive, feed and mate normally (Picciotto et al. 1995). Their brains have a normal size and morphology. The high-affinity nicotine-binding sites (classically attributed to $\alpha 4\beta 2$ nAChRs) completely disappear from the brain of homozygous mutant mice, whereas the α-bungarotoxin sites (corresponding to the $\alpha 7$-containing nAChRs) persist. Electrophysiological responses to nicotine are no longer recorded in the thalamus but persist in a few structures expressing the $\beta 4$ nAChR subunit (such as the medial habenula). Further analysis of the $\beta 2$ mutant mice has shown that the $\beta 2$-containing nAChRs are expressed both in the somatodendritic compartment and in the axonal compartment of neurons as presynaptic nAChRs (Léna and Changeux 1997). The absence of the $\beta 2$ subunit affects the performance of mutant animals on associative memory (passive avoidance) tests and suppresses the improvement of the performance by nicotine (Picciotto et al. 1995). Activation of $\beta 2$-containing nAChRs by endogenous acetylcholine probably takes place during the course of these memory tasks.

Preliminary results indicate that knock-out of the α7 subunit yields animals that survive normally but display an anomalous synchronisation on electroencephalograph (EEG) recordings (Orr Urtreger et al. 1996).

Nicotine enhancement of memory processes has motivated clinical trials of nicotinic treatment in Alzheimer's disease (AD). The severity of symptoms in AD is well correlated with a reduction in cortical acetylcholine (see Bierer et al. 1995 and references therein), and AD patients exhibit a marked reduction in the number of high-affinity nicotine binding sites (Perry et al. 1995). Nicotine treatment partially relieves the cognitive deficits of AD (Newhouse et al. 1988; Jones et al. 1992). The site of this beneficial action of nicotine is not yet clearly established. For instance, nicotine may increase the levels of acetylcholine in the cortex by recruiting presynaptic nAChRs on acetylcholine terminals in the cortex (e.g. Marchi and Raiteri 1996). As the use of nicotine presents a number of side effects linked to the activation of peripheral nAChRs, attempts have been made to find nicotinic drugs specific for brain subtypes, such as ABT418 (Arneric et al. 1995), SIB-1508Y (Cosford et al. 1996) and RJR-2403 (Lipiello et al. 1996).

Nicotinic Receptors in the Mesostriatal Reward System and Tobacco Abuse

The nAChRs subunits are expressed abundantly in the mesencephalic dopaminergic nuclei (see Le Novère et al. 1996 and refernces therein). These nuclei are part of the mesostriatal reward system. Theoretical work has underlined the critical function of reward systems in learning by selection of behavioural rules (Dehaene and Changeux 1991; Pennartz 1996). Dysfunction or anomalous chemical stimulation of these systems strongly affects brain function. Indeed, the mesostriatal dopaminergic system is a common target of many addictive drugs (reviewed in Changeux et al. 1996; Altman et al. 1996).

Accumulating data suggest that both tobacco smoking in humans and nicotine self-administration in animals are associated with an increase in dopamine release following nicotinic actions on mesencephalic dopaminergic neurons (reviewed in Rose and Corrigall 1997). Self-administration of nicotine shares common mechanisms with that of other addictive drugs. Minimal doses of nicotine, comparable to those producing self-administration behaviour, trigger a specific increase of metabolism and release of dopamine in the nucleus accumbens, as is observed for strongly addictive drugs such as cocaine and amphetamines (Pontieri et al. 1996). Nicotine and cocaine self-administration activates a number of common brain structures, as visualized with cFos immunoreactivity to reveal neuronal activation, notably the terminal fields to the mesencephalic dopaminergic neurons (Merlo et al. 1997). It should be noted that tobacco smoking is not only associated with nicotine intake but also with respiratory sensations of smoke intake (reviewed in Rose and Carrigall 1997); nicotine action could also be amplified by changes in dopamine metabolism, as smokers display a 40% reduction of monoamine oxidase B compared to former smokers and nonsmokers (Fowler et al. 1996).

What is the composition of the nAChRs involved in the self-administration of nicotine? The concentration of nicotine in the plasma of smokers is in the 100–500 nM range (Henningfield et al. 1983). Oocyte experiments with human nAChRs reveal which combinations of subunits may respond to such low concentration of agonists *in vivo*. EC_{50} values below 10 µM have been observed for $\alpha4\beta2$, $\alpha4\beta4$ (Chavez Noriega et al. 1997), $\alpha3\beta2$ (Gernazich et al. 1995, but see Chavez Noriega 1997) and $\alpha3\beta2\alpha5$ (Wang et al. 1996). *In situ* hybridization experiments (references in Le Novère et al. 1996) indicate that the nAChR subunits forming these combinations (except $\beta4$) are expressed in the mesencephalic nuclei. These nuclei also contain high amounts of the $\alpha6$ and $\beta3$ subunit mRNAs, suggesting the contribution of an $\alpha6\beta3\beta2$ subtype (Le Novère et al. 1996). The functional significance of this combination of nAChR subunits awaits demonstration, for instance, in a reconstituted expression system. Recent data for the $\beta2$ knock-out mice indicate that $\beta2$ is part of the nAChRs involved in nicotine reinforcement (Picciotto et al. 1998)]. The knock-out of other nAChR subunits will be necessary to identify fully the composition of the main nAChR subtype(s) in the mesencephalic dopaminergic neurons.

The Nicotinic Receptors in Psychiatric and Neurological Disorders

The high prevalence of tobacco smoking in schizophrenic patients suggests nicotine intake by cigarette consumption may be a form of self-medication. By stimulating the mesencephalic dopaminergic system (see above), more specifically by increasing the burst firing of dopaminergic neurons, nicotine might compensate for the hypofrontality observed in schizophrenia (Nisell et al. 1995). Nicotine has been found to reverse the cognitive deficits produced by haloperidol in schizophrenics (Levin et al. 1996 a). A synergy between nicotine and dopaminergic neuroleptics also exists in the treatment of Tourette's syndrome (Shytle et al. 1996). Nicotine has been proposed as an alternative to drugs increasing the brain levels of dopamine in the treatment of attention deficit/hyperactivity disorder (Levin et al. 1996 b). Finally, nicotine and various nicotinic agonist might help to compensate the deficit in striatal dopamine in Parkinson's disease patients and might, in some instances, relieve the symptom of the disease (Fagerstrom et al. 1994, but see Clemens et al. 1995). The interaction of nicotine with the dopaminergic system might thus explain its efficacy in the treatment of psychiatric and neurological disorders.

Schizophrenic patients often exhibit a diminished habituation to auditory stimulation among diverse symptoms (reviewed in Leonard et al. 1996). Experiments in the rodent have shown that auditory gating is impaired by antagonists of $\alpha7$ nAChRs. On the other hand, the number of α-bungarotoxin sites is reduced in post-mortem brains of schizophrenics (Freedman et al. 1995). The deficit of sensory gating in schizophrenics might thus be attributable to a reduction or a loss of $\alpha7$ nAChR function. Consistent with this hypothesis, genetic analysis in nuclear families with at least two cases of schizophrenia has shown that the deficit in auditory gating is significantly linked to a genetic marker neighbouring the locus of the α7 gene (Freedman et al. 1997). A study of the relatives of schizophrenics sharing the deficit in auditory gating revealed that nicotine could reverse the deficit, presumably by activating α7 nAChRs (references in Leonard et al. 1996). This result is quite unexpected, as human α7 nAChRs exhibit a low sensitivity to nicotine (see above). However, recent experiments in the chick (McGehee et al. 1995) and rat (Gray et al. 1996) have shown that low doses of nicotine can activate α-bungarotoxin-sensitive nAChRs in glutamatergic nerve terminals. The α7 nAChR subunit, possibly associated with still unidentified subunit(s), may thus form another relevant target for nicotinic therapies of psychiatric disorders.

Conclusions

Over 25 years after the identification and purification of the electric fish nAChR (see Changeux 1980), the accumulating knowledge of the nAChRs in vertebrates has led to the demonstration that alterations of these receptors are responsible for a variety of familial disorders of the central and peripheral nervous system. Conversely, these receptors are now considered as relevant targets for nicotinic therapies of brain disorders.

Previous experiments combining photoaffinity labelling and site-directed mutagenesis had shown that changes of critical amino acids in nAChR channel or ligand-binding sites may markedly affect its function in a pleiotropic manner, and may, for instance, either reduce or increase channel opening in the presence and sometimes in the absence of acetylcholine by altering the allosteric properties of the protein. Interestingly, analogous, if not identical, point mutations in human nAChR genes (and glycine receptor genes) have been shown to cause pathologies either by a loss or by a increase of function. Mutations causing pathologies via changes in allosteric properties have also been described for G-protein-linked receptors (Lefkowitz et al. 1993). Development of novel nicotinic therapies with pharmacological agents targeted to these diverse 'allosteric' phenotypes may thus be anticipated.

While the strategic location of nAChRs in the dopaminergic reward system renders nicotine an additive drug, it also underlies potential beneficial effects of nicotine in the treatment of psychiatric disorders. Furthermore, nAChRs may relieve symptoms of AD or schizophrenia via pathways different from the dopaminergic system. Therefore, nicotinic agents specifically activating nAChR subtypes absent from the dopaminergic system, and thus with no (or less) addictive properties, should be sought.

Fundamental research on the properties of nAChRs in normal and pathological situations opens many new strategies to design drug therapies targeted not only to specific nAChRs in defined brain circuits but also to specific allosteric transitions impaired by nAChR gene mutations in humans.

Acknowledgements

We thank A. Devillers-Thiéry, S. Edelstein, N. Le Novère, R. Miles, M. Picciotto and M. Zoli for critical reading of the manuscript, and Y. Paas and N. Le Novère for their help with the table and figures. This work was supported by grants from the Collège de France, the Centre National de la Recherche Scientifique, the Association Française contre les Myopathies, Direction des Recherches et Etudes Techniques, Tobacco Research and the EEC Biotech and Biomed Programs. C. Léna is supported by the Institut Pasteur.

References

Altman J, Everitt BJ, Glautier S, Markou N, Nutt D, Oretti R, Phillips GD, Robbins TW (1996) The biological, social and clinical bases of drug addiction: commentary and debate. Psychopharmacology 125:285–345.

Arneric SP, Sullivan JP, Decker MW, Brioni JD, Bannon AW, Briggs CA, Donnelly RD, Radek RJ, Marsh KC, Kyncl J et al. (1995) Potential treatment of Alzheimer disease using cholinergic channel activators (ChCAs) with cognitive enhancement, anxiolytic-like, and cytoprotective properties. Alzheimer Dis Assoc Disord 9 (suppl 2):50–61.

Bertrand D, Devillers-Thiery A, Revah F, Galzi JL, Hussy N, Mulle C, Bertrand S, Ballivet M, Changeux JP (1992) Unconventional pharmacology of a neuronal nicotinic receptor mutated in the channel domain. Proc Natl Acad Sci USA 89:1261–1265.

Bertrand D, Galzi JL, Devillers-Thiery A, Bertrand S, Changeux JP (1993) Mutations at two distinct sites within the channel domain M2 alter calcium permeability of neuronal $\alpha7$ nicotinic receptor. Proc Natl Acad Sci USA 90:6971–6975.

Bierer LM, Haroutunian V, Gabriel S, Knott PJ, Carlin LS, Purohit DP, Perl DP, Schmeidler J, Kanof P, Davis KL (1995) Neurochemical correlates of dementia severity in Alzheimer's disease: relative importance of the cholinergic deficits. J Neurochem 64:749–760.

Buisson B, Gopalakrishnan M, Arneric SP, Sullivan JP, Bertrand D (1996) Human $\alpha4\beta2$ neuronal acetylcholine receptor in HEK 293 cells: a patch-clamp study. J Neurosci 16:7880–7891.

Changeux JP (1980) The acetylcholine receptor: an "allosteric" membrane protein. Harvey Lectures 75:85–254.

Changeux JP (1990) Functional architecture and dynamics of the nicotinic acetylcholine receptor: an allosteric ligand-gated ion channel. In: Llinas RR, Changeux JP, Purves L, Bloom FE (eds), Fidia Research Foundation Neuroscience Award Lectures. Raven Press, New York, pp 21–168.

Changeux JP, Bessis A, Bourgeois JP, Corringer PJ, Devillers-Thiéry A, Eisele JL, Kerszberg M, Lena C, Le Novere N, Picciotto M, Zoli M (1996) Nicotinic receptors and brain plasticity. Cold Spring Harb Symp Quant Biol 61:343–362.

Chavez Noriega LE, Crona JH, Washburn MS, Urrutia A, Elliott KJ, Johnson EC (1997) Pharmacological characterization of recombinant human neuronal nicotinic acetylcholine receptors h-$\alpha2\beta2$, h-$\alpha2\beta4$, h-$\alpha3\beta2$, h-$\alpha3\beta4$, h-$\alpha4\beta2$, h-$\alpha4\beta4$ and h-$\alpha7$ expressed in *Xenopus oocytes*. J Pharmacol Exp Ther 280:346–356.

Clemens P, Baron JA, Coffey D, Reeves A (1995) The short-term effect of nicotine chewing gum in patients with Parkinson's disease. Psychopharmacology 117:253–256.

Cosford ND, Bleichert L, Herbaut A, McCallum JS, Vernier JM, Dawson H, Whitten JP, Adams P, Chavez NL, Correa LD et al. (1996) (S)-(-)-5-ethynyl-3-(1-methyl-2-pyrrolidinyl)pyridine maleate (SIB-1508Y): a novel anti-parkinsonian agent with selectivity for neuronal nicotinic acetylcholine receptors. J Med Chem 39:3235–3237.

Croxen R, Newland C, Beeson D, Oosterhuis H, Chauplannaz G, Vincent A, Newsom-Davis J (1997) Mutations in different functional domains of the human muscle acetylcholine receptor α subunit in patients with the slow-channel congenital myasthenic syndrome. Hum Mol Genet 6:767–774.

Dehaene S, Changeux JP (1991) The Wisconsin card sorting test: theoretical analysis and modeling in a neuronal network. Cereb Cortex 1:62–79.

Devillers-Thiery A, Galzi JL, Bertrand S, Changeux JP, Bertrand D (1992) Stratified organization of the nicotinic acetylcholine receptor channel. Neuroreport 3:1001–1004.

Edelstein SJ, Changeux JP (1996) Allosteric proteins after thirty years: the binding and state functions of the neuronal $\alpha 7$ nicotinic acetylcholine receptors. Experientia 52:1083–1090.

Edelstein SJ, Schaad O, Henry E, Bertrand D, Changeux JP (1996) A kinetic mechanism for nicotinic acetylcholine receptors based on multiple allosteric transitions. Biol Cybern 75:361–379.

Engel AG, Ohno K, Bouzat C, Sine SM, Griggs RC (1996 a) End-plate acetylcholine receptor deficiency due to nonsense mutations in the ε subunit. Ann Neurol 40:810–817

Engel AG, Ohno K, Milone M, Wang HL, Nakano S, Bouzat C, Pruitl J, Hutchinson DO, Brengman JM, Bren N, Sieb JP, Sine SM (1996 b) New mutations in acetylcholine receptor subunit genes reveal heterogeneity in the slow-channel congenital myasthenic syndrome. Hum Mol Genet 5:1217–1227.

Fagerstrom KO, Pomerleau O, Giordani B, Stelson F (1994) Nicotine may relieve symptoms of Parkinson's disease. Psychopharmacology 116:117–119.

Fowler JS, Volkow ND, Wang GJ, Pappas N, Logan J, MacGregor R, Alexoff D, Shea C, Schlyer D, Wolf AP et al. (1996) Inhibition of monoamine oxidase B in the brains of smokers. Nature 379:733–736.

Freedman R, Hall M, Adler LE, Leonard S (1995) Evidence in postmortem brain tissue for decreased numbers of hippocampal nicotinic receptors in schizophrenia. Biol Psychiatry 18:537–551.

Freedman R, Coon H, Mylesworsley M, Orr Urtreger A, Olincy A, Davis A, Polymeropoulos M, Holik J, Hopkins J, Hoff M et al. (1997) Linkage of a neurophysiological deficit in schizophrenia to a chromosome 15 locus. Proc Natl Acad Sci USA 94:587–592.

Galzi JL, Bertrand D, Devillers-Thiery A, Revah F, Bertrand S, Changeux JP (1991) Functional significance of aromatic amino acids from three peptide loops of the $\alpha 7$ neuronal nicotinic receptor site investigated by site-directed mutagenesis. FEBS Lett 294:198–202.

Galzi JL, Changeux JP (1995) Neuronal nicotinic receptors: molecular organization and regulations. Neuropharmacology 34:563–582.

Galzi JL, Edelstein SJ, Changeux J (1996) The multiple phenotypes of allosteric receptor mutants. Proc Natl Acad Sci USA 93:1853–1858.

Gerzanich V, Peng X, Wang F, Wells G, Anand R, Fletcher S, Lindstrom J (1995) Comparative pharmacology of epibatidine: a potent agonist for neuronal nicotinic acetylcholine receptors. Mol Pharmacol 48:774–782.

Gerzanich V, Kuryatov A, Anand R, Lindstrom J (1997) Orphan $\alpha 6$ nicotinic AChR subunit can from a functional heteromeric acetylcholine receptor. Mol Pharmacol 51:320–327.

Gomez CM, Gammack JT (1995) A leucine-to-phenylalanine substitution in the acetylcholine receptor ion channel in a family with the slow-channel syndrome. Neurology 45:982–985.

Gomez CM, Maselli R, Gammack J, Lasalde J, Tamamizu S, Cornblath DR, Lehar M, McNamee M, Kuncl RW (1996) A β-subunit mutation on the acetylcholine receptor channel gate causes severe slow-channel syndrome. Ann Neurol 39:712–723.

Gomez CM, Maselli R, Gundeck JE, Chao M, Day JW, Tamamizu S, Lasalde JA, McNamee M, Wollmann RL (1997) Slow channel transgenic mice – a model of postsynaptic organellar degeneration at the neuromuscular junction. J Neurosci 17:4170–4179.

Gopalakrishnan M, Buisson B, Tourna E, Giordano T, Campbell JE, Hu IC, Donnelly RD, Arneric SP, Bertrand D, Sullivan JP (1995) Stable expression and pharmacological properties of the human alpha7 nicotinic acetylcholine receptor. Eur J Pharmacol 290:237–246.

Gray R, Rajan AS, Radcliffe KA, Yakehiro M, Dani JA (1998) Hippocampal synaptic transmission enhanced by low concentrations of nicotine. Nature 383:713–716.

Henningfield JE, Miyasato K, Jasinski DR (1983) Cigarette smokers self-administer intravenous nicotine. Pharmacol Biochem Behav 19:887–890.

Jones GM, Sahakian BJ, Levy R, Warburton DM, Gray JA (1992) Effects of acute subcutaneous nicotine on attention, information processing and short-term memory in Alzheimer's disease. Psychopharmacology 108:485–494.

Labarca C, Nowak MW, Zhang H, Tang L, Deshpande P, Lester HA (1995) Channel gating governed symmetrically by conserved leucine residues in the M2 domain of nicotinic receptors. Nature 376:514–516.

Le Novère N, Zoli M, Changeux JP (1996) Neuronal nicotinic receptor α6 subunit mRNA is selectively concentrated in catecholaminergic nuclei of the rat brain. Eur J Neurosci 8:2428–2439.

Lefkowitz RJ, Cotecchia S, Samama P, Costa T (1993) Constitutive activity of receptors coupled to guanine nucleotide regulatory proteins. Trends Pharmacol Sci 14:303–307.

Léna C, Changeux JP (1997) Role of Ca^{2+} ions in nicotinic facilitation of GABA release in mouse thalamus. J Neurosci 17:576–585.

Leonard S, Adams C, Breese CR, Adler LE, Bickford P, Byerley W, Coon H, Griffith JM, Miller C, Myles WM et al. (1996) Nicotinic receptor function in schizophrenia. Schizophr Bull 22:431–445.

Levin ED, Wilson W, Rose JE, McEvoy J (1996 a) Nicotine-haloperidol interactions and cognitive performance in schizophrenics. Neuropsychopharmacology 15:429–436.

Levin ED, Conners CK, Sparrow E, Hinton SC, Erhardt D, Meck WH, Rose JE, March J (1996 b) Nicotine effects in adults with attention-deficit/hyperactivity disorder. Psychopharmacology 123:55–63.

Lippiello PM, Bencherif M, Gray JA, Peters S, Grigoryan G, Hodges H, Collins AC (1996) RJR-2403: a nicotinic agonist with CNS selectivity II. *In vivo* characterization. J Pharmacol Exp Ther 279:1422–I1429.

Lynch JW, Rajendra S, Pierce KD, Handford CA, Barry PH, Schofield PR (1997) Identification of intracellular and extracellular domains mediating signal transduction in the inhibitory glycine receptor chloride channel. EMBO J 16:110–120.

Marchi M, Raiteri M (1996) Nicotinic autoreceptors mediating enhancement of acetylcholine release become operative in conditions of 'impaired' cholinergic presynaptic function. J Neurochem. 67:1974–1981.

McGehee DS, Role LW (1995) Physiological diversity of nicotinic acetylcholine receptors expressed by vertebrate neurons. Annu Rev Physiol 57:521–546.

McGehee DS, Heath MJ, Gelber S, Devay P, Role LW (1995) Nicotine enhancement of fast excitatory synaptic transmission in CNS by presynaptic receptors. Science 269:1692–1696.

Merlo Pich E, Pagliusi SR, Tessari M, Talabot-Ayer D, Hooft van Huijsduijnen R, Chiamulera C (1997) Common neural substrates for the addictive properties of nicotine and cocaine. Science 275:83–86.

Milone M, Wang H, Ohno K, Fukudome T, Pruitt J, Bren N, Sine S, Engel A (1997) Slow-channel myasthenic syndrome caused by enhanced activation, desensitization and agonist binding affinity attributable to mutation in the M2 domain of acetylcholine receptor α subunit. J Neurosci 17:5651–5665.

Newhouse PA, Sunderland T, Tariot PN, Blumhardt CL, Weingartner H, Mellow A, Murphy DL (1988) Intravenous nicotine in Alzheimer's disease: a pilot study. Psychopharmacology 95:171–175.

Nisell M, Nomikos GG, Svensson TH (1995) Nicotine dependence, midbrain dopamine systems and psychiatric disorders. Pharmacol Toxicol 76:157–162.

Ohno K, Hutchinson DO, Milone M, Brengman JM, Bouzat C, Sine SM, Engel AG (1995) Congenital myasthenic syndrome caused by prolonged acetylcholine receptor channel openings due to a mutation in the M2 domain of the ε subunit. Proc Natl Acad Sci USA 92:758–762.

Ohno K, Wang HL, Milone M, Bren N, Brengman JM, Nakano S, Quiram P, Pruitt JN, Sine SM, Engel AG (1996) Congenital myasthenic syndrome caused by decreased agonist binding affinity due to a mutation in the acetylcholine receptor ε subunit. Neuron 17:157–170.

Ohno K, Quiram PA, Milone M, Wang HL, Harper MC, Pruitt JN, Brengman JM, Pao L, Fischbeck KH, Crawford TO et al. (1997) Congenital myasthenic syndromes due to heteroallelic nonsense/missense mutations in the acetylcholine receptor ε subunit gene-identification and functional characterization of six new mutations. Hum Mol Genet 6:753–766.

Orr Urtreger A, Noebels JL, Goldner FM, Patrick J, Beaudet AL (1996) A novel hypersynchronous neocortical EEG phenotype in mice deficient in the neuronal nicotinic acetylcholine receptor (nAChRs) α7 subunit gene. Am J Hum Genet 59:A53.

Peng X, Katz M, Gerzanich V, Anand R, Lindstrom J (1994) Human α7 acetylcholine receptor: cloning of the α7 subunit from the SH-SY5Y cell line and determination of pharmacological properties of native receptors and functional α7 homomers expressed in *Xenopus oocytes*. Mol Pharmacol 45:546–554.

Pennartz C (1996) The ascending neuromodulatory systems in learning by reinforcement-comparing computational conjectures with experimental findings. Brain Res Rev 21:219–245.

Perry EK, Morris CM, Court JA, Cheng A, Fairbairn AF, McKeith IG, Irving D, Brown A, Perry RH (1995) Alteration in nicotine binding sites in Parkinson's disease, Lewy body dementia and Alzheimer's disease: possible index of early neuropathology. Neuroscience 64:385–395.

Picciotto MR, Zoli M, Léna C, Bessis A, Lallemand Y, Le Novère N, Vincent P, Merlo Pich E, Brûlet P, Changeux JP (1995) Abnormal avoidance learning in mice lacking functional high-affinity nicotine receptor in the brain. Nature 374:65–67.

Picciotto MR, Zoli M, Rimondini R, Léna C, Marubio L, Merlo Pich E, Fuxe K, Changeux JP (1998) Acetylcholine receptors containing the β2 subunit are involved in the reinforcing properties of nicotine. Nature 391:173–177.

Pontieri FE, Tanda G, Orzi F, Di Chiara G (1996) Effects of nicotine on the nucleus accumbens and similarity to those of addictive drugs. Nature 382:255–257.

Ramirez Latorre J, Yu CR, Qu X, Perin F, Karlin A, Role L (1996) Functional contribution of α5 subunit to neuronal acetylcholine receptor channels. Nature 380:347–351.

Revah F, Bertrand D, Galzi JL, Devillers-Thiery A, Mulle C, Hussy N, Bertrand S, Ballivet M, Changeux JP (1991) Mutations in the channel domain alter desensitization of a neuronal nicotinic receptor. Nature 353:846–649.

Role LW, Berg DK (1996) Nicotinic receptors in the development and modulation of CNS synapses. Neuron 16:1077–1085.

Rose JE, Corrigall WA (1997) Nicotine self-administration in animals and humans-similarities and differences. Psychopharmacology 130:28–40.

Shytle RD, Silver AA, Philipp MK, McConville BJ, Sanberg PR (1996) Transdermal nicotine for Tourette's syndrome. Drug Dev Res 38:290–298.

Sine SM, Ohno K, Bouzat C, Auerbach A, Milone M, Pruitt JN, Engel AG (1995) Mutation of the acetylcholine receptor α subunit causes a slow-channel myasthenic syndrome by enhancing agonist binding affinity. Neuron 15:229–239.

Steinlein OK, Mulley JC, Propping P, Wallace RH, Phillips HA, Sutherland GR, Scheffer IE, Berkovic SF (1995) A missense mutation in the neuronal nicotinic acetylcholine receptor α4 subunit is associated with autosomal dominant nocturnal frontal lobe epilepsy. Nat Genet 11:201–203.

Steinlein OK, Magnusson A, Stoodt J, Bertrand S, Weiland S, Berkovic SF, Nakken KO, Propping P, Bertrand D (1997) An insertion mutation of the CHRNA4 gene in a family with autosomal dominant nocturnal frontal lobe epilepsy. Hum Mol Genet 6:943–947.

Treining M, Chalfie M (1995) A mutated acetylcholine receptor subunit causes neuronal degeneration in C. elegans. Neuron 14:871–877.

Tsigelny I, Sugiyama N, Sine SM, Taylor P (1997) A model of the nicotinic receptor extracellular domain based in sequence identity and residue location. Biophys J 73:52–66.

Wang F, Gerzanich V, Wells GB, Anand R, Peng X, Keyser K, Lindstrom J (1996) Assembly of human neuronal nicotinic receptor α5 subunits with α3, β2, and β4 subunits. J Biol Chem 271:17656–17665.

Weiland S, Witzemann V, Villarroel A, Propping P, Steinlein O (1996) An amino acid exchange in the second transmembrane segment of a neuronal nicotinic receptor causes partial epilepsy by altering its desensitization kinetics. FEBS Lett 398:91–96.

Witzemann V, Schwarz H, Koenen M, Berberich C, Villarroel A, Wernig A, Brenner HR, Sakmann B (1996) Acetylcholine receptor ε-subunit deletion causes muscle weakness and atrophy in juvenile and adult mice. Proc Natl Acad Sci USA 93:13286–13291.

Zoli M, Le Novère N, Hill JJ, Changeux JP (1995) Development regulation of nicotinic ACh receptor subunit mRNAs in the rat central and peripheral nervous systems. J Neurosci 15:1912–1939.

Towards an Ecology of Cortical Organization: Experience and the Changing Brain

B. Kolb[*]

Summary

Evidence is accumulating to show that the circuitry of the mammalian cortex is reorganized by experience throughout the lifetime of an individual. This phenomenon is known as *plasticity*. There are multiple mechanisms of plasticity that range from gross cortical changes such as the generation of neurons and glia, to more subtle changes such as the alteration of synapses or changes in the production of chemical messengers. Experiences that affect brain morphology include a broad category of events including the effect of sensory events, gonadal hormones, aging, brain injury, stress, and drugs and other chemicals. We now have evidence that each of these experiences can alter cortical organization and they do so in remarkably similar ways. One important characteristic of plasticity is that its nature varies with age. The cortex is most plastic at a critical time in development, which is during the time of dendritic and synaptic growth and the generation of astrocytes. In humans, this period begins shortly after birth and continues for about two years. It is possible to influence the extent of plasticity later in life, such as in adulthood, by recreating the conditions that were present in the brain during the critical period in development. This can be done either with the addition of chemicals, such as neurotrophins, or by behavioral manipulations. An important principle of cortical plasticity is that modifications of cortical structure are reflected in behavioral change. Behavioral change may be referred to by many names, including learning, maturation, recovery, addiction, etc. In sum, the capacity of the cortex to change is constrained by the life history of the individual or, stated differently, by the ecology that brain finds itself in throughout its lifetime.

Introduction

One of the most intriguing questions in behavioral neuroscience concerns the manner in which the brain, and especially the neocortex, can modify its structure and ultimately its function throughout one's lifetime. The idea that sensory

[*] Bryan Kolb, Dept of Psychology and Neuroscience University of Lethbridge, Lethbridge, AB, Canada, T1K 3M4, PH: 403-329-2405, FAX: 403-329-2555, E-mail: Kolb@uleth.ca

J. Grafman / Y. Christen (Eds.)
Neuronal Plasticity:
Building a Bridge from the Laboratory to the Clinic
© Springer-Verlag Berlin Heidelberg New York 1999

experience changes brain structure has a long history. Almost 100 years ago Ramon y Cajal suggested that learning could produce prolonged morphological changes in neurons. In 1948 Konorski hypothesized that the morphological changes would be activity-dependent and, in 1949, Hebb proposed that the critical changes would be synaptic. Thus, long-term behavioral changes, which occur as a result of learning, were hypothesized to be supported by changes in the synaptic organization of the nervous system. Over the past 50 years considerable evidence has accumulated to support this idea. For example, Greenough and his colleagues have shown that experience increases the extent of dendritic arbor in cortical pyramidal cells and that this dendritic growth is associated with an increase in the number of synapses per neuron (for a review, see Greenough and Chang 1989). These synaptic changes are associated with behavioral changes, including learning.

Recently, however, it has become clear that there is more to the story than Cajal, Konorski, or Hebb could have imagined. Experience-dependent changes in the brain are far more extensive than just synaptic change. Furthermore, the definition of experience has expanded considerably to include drug and hormone experience, brain injury, and aging (e. g., Kolb 1995). Thus, it is now becoming evident that the brain's structure is subject to continual change in general morphology of both its fundamental unit (the neuron) and glial cells and changes in various chemical messengers related to both neuronal and glial function. It is becoming apparent, therefore, that an understanding of the brain and its relation to behavior requires an understanding of the brain in its environment, which we can describe as an *ecological analysis of the brain*. Ecology comes from the Greek word "oikos," meaning "household" or "place to live". Thus, an ecology of the brain deals with the brain in its environment, both internal and external. Although the concept of environmental sculpting presumably applies to all of the brain, my own research is restricted largely to the cerebral cortex. Thus, the goal of this chapter is to develop what could be called an Ecological Theory of Cortical Organization. The basic premise of such a theory is that the structure, and ultimately the function, of the cortex is a product of its environmental history over its lifetime. Thus, to understand the cortex and its operations we must understand how the environment constructs the brain throughout our lifetime. I hasten to point out that the environment is not operating upon an unstructured brain because there is a genetic template that determines the general organization of the cortex. Rather, the environment can be conceived as a set of chisels that restructures the genetic template.

In the current chapter I shall focus on four lines of inquiry that have the following objectives: 1) to identify the age-dependent differences in brain plasticity; 2) to identify the nature of plastic changes after cortical injury; 3) to identify ways to restore or increase brain plasticity in both the normal and injured brain; and 4) to identify similarities between drug-induced plasticity and change resulting from sensory experience or injury.

How Does Plasticity Change with Age?

The basic postulate here is that the mechanisms of plasticity available to the brain vary with age. Thus, the effects of experience on the brain will vary with age. This, of course, conforms to the common view that the infant brain is more labile than the adult brain. It is not so simple as it first appears, however. There are times in early life when the brain appears especially plastic and other times when it appears far less plastic than even in senescence. To understand these differences we must first consider briefly the development of the brain.

Neuronal Changes during Development

The mammalian brain follows a general pattern of development, beginning as a hollow tube in which a thin layer of presumptive neural cells surrounds a single ventricle. The development of the brain from the neural tube involves several stages including: cell birth (mitosis), cell migration, cell differentiation, dendritic and axonal growth, synaptogenesis, and cell death and synaptic pruning (Fig. 1). The order of these events is similar across species, but because the gestation time varies dramatically across different mammalian species, the timing of the events relative to birth varies considerably. This can be seen in the common observation that whereas kittens and puppies are born helpless and blind (their eyes do not open for about two weeks), human babies are born somewhat more mobile and with their eyes open, and calves at birth are able to stand and walk about and, of course, have their eyes open. It is worth noting here that rats, which are the subject of choice in most plasticity and recovery studies, are born even less mature than kittens and their eyes do not open until about postnatal day 15. They are weaned around 21 days of age, reach adolescence about 60 days of age, and can be considered adults by about 90 days of age.

In the rat, neuron birth in the cortex begins about embryonic day 12 (E12) and continues until about E21 (e.g., Uylings et al. 1990). (Birth occurs on about E22.) Neuron migration begins shortly after cell birth and continues in the cortex until about postnatal day 6 (P6). The development of dendrites, axons, and synapses begins once cells arrive at their final destination and differentiate into a particular cell type. The peak rate of this growth is probably around P10–15, although it continues for some time afterward. Glial development occurs later than neural development, with astrocyte growth reaching its peak in the cortex around P7–10.

As we look at cortical development, we can identify several key elements that will contribute to cortical plasticity. First, there is the genesis of neurons. Although the genesis of cortical neurons is normally complete by birth in the rat, neurons are generated postnatally, and throughout life, for the olfactory bulb and dentate gyrus of the hippocampus. Furthermore, the stem cells for neurons remain active in the subventricular zone throughout life. This implies that *neurogenesis is possible for the cortex throughout life*. The trick is to figure out the

20   B. Kolb

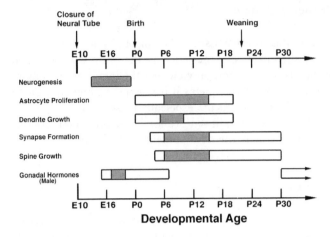

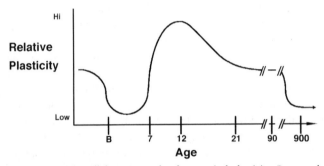

Fig. 1. Top. Main cellular events related to cortical plasticity. Bars mark the approximate beginning and ending of different processes. The intensity of the shading reflects the intensity of the phenomenon. **Bottom.** Summary of the time-dependent differences in cortical plasticity

"switch" to control it. (I will return to this below.) Second, there is the genesis of glial cells, especially astrocytes. Astrocytes play a special role in plasticity, as they manufacture various chemical messengers, including some neurotrophins. As a result, an increase in astrocyte production will facilitate plasticity; an absence of astrocytes, such as during the first days after birth, will retard plasticity. Third, there is the production of synapses, which likely requires dendritic changes. This can occur throughout life but the most active time is from about P7–P15, which implies that this might be an especially plastic time for the cortex. Finally, there is the cell death and synaptic pruning, which provides a mechanism for fine tuning the connectivity of the cortex. This is likely most active from about P15–P30, although it will continue throughout life.

Taking these elements together, we can make some predictions. First, the period immediately following birth (say P1–P6) is likely to be a time of limited plasticity because there are few cortical astrocytes. There is also limited synapse formation. Second, the period from P7–P15 may be a time of maximal plasticity because the cortex is actively making connections and astrocytic activity is high. Third, the period from about P15–P30 may be more plastic than later in life because there is the unique period of neuron death and pruning. One could imagine that experience would influence the rate, and perhaps the extent, of neuron death. Next, as rats reach puberty around P60, we can imagine that there will be a special period of plasticity, as gonadal hormones are known to influence cell structure and connectivity. Finally, after puberty we might see a slow diminution of plasticity as the brain ages.

It is difficult, and perhaps even hazardous, to try to identify a precise human analogue to the plastic phases in the life of the rat. Nonetheless, we can make a general case that the plastic embryonic period will be sometime in the second trimester of gestation. As neurogenesis ends in the third trimester, this is probably equivalent to the period of poor plasticity in newborn rats. It is uncertain just how long this period continues, but on the basis of the effects of birth injury on subsequent brain functioning, it seems likely that it includes the early postnatal period. The human brain then enters a period of maximum dendritic and synaptic growth in the cortex, continuing until somewhere around 2 years. This marks the highly plastic period. After age 2 there is a gradual decline in plasticity until adulthood. As in rats, there is an adolescent period that is characterized by remodeling of circuits by gonadal hormones. Finally, in senescence, there is a rapid decline in plasticity.

Age and the Changing Brain

One way to investigate age-dependent change is to expose animals to specific experiences at different times in their life (for a review, see Kolb et al. 1998). For example, in one series of experiments we placed rats in "enriched environments" for three months beginning at weaning or in young adulthood (Fig. 2). The animals were later studied in an extensive battery of behavioral tests and compared to littermate controls who were housed in standard laboratory cages. The principal result was that enriched experience facilitated behavioral performance at both ages but there were very different effects upon dendritic structure from the experience at the two ages (Fig. 3). Animals placed in the enriched conditions in young adulthood showed a large *increase* in dendritic arborization as well as an *increase* in spine density of cortical pyramidal neurons (Fig. 4). In contrast, however, animals placed in the enriched environments as juveniles (at weaning) showed an *increase* in branching and a *decrease* in spine density. These results imply that there is a qualitatively different structural change in pyramidal neurons at different ages. The effect in the juvenile animals was intriguing because the spine density decrease implied a decrease in overall synapse number per neuron relative to animals with the same treatment in adulthood. This result is hard to reconcile with their enhanced behavioral capacities relative to the older animals. We therefore decided to investigate the effects of experience earlier in life.

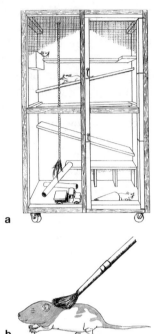

a

b

Fig. 2. A. Schematic illustration of the rat condominiums used in studies of the effects of enriched experience. **B.** Schematic illustration of the tactile-stroking paradigm

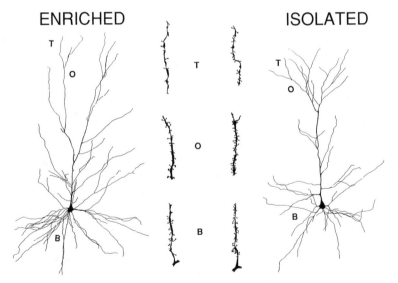

ENRICHED

ISOLATED

Fig. 3. Illustration of representative layer III parietal pyramidal neurons from a rat placed in an enriched environment at weaning versus a littermate that was housed in standard laboratory housing. The dendritic branches down the midline are expanded view of terminal (T), oblique (O), and basilar (B) portions, illustrating the dendritic spines. Note that the spine density varies with location on the dendritic tree. Enriched housing at this age produced an increase in branching but a decrease in spine density

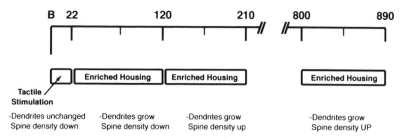

Fig. 4. A schematic illustration of the effects of sensory experience at different developmental times. The effect of experience varies qualitatively and quantitatively with age

We gave newborn animals two weeks of tactile stimulation (brushing with a small brush for 15 min 3X per day). This experience had *no effect* on dendritic morphology in adulthood but led to a *decrease* in spine density that was present by 21 days of age and persisted into adulthood. Curiously, animals with this early tactile experience showed better performance in skilled motor tasks and in the learning of spatial mazes than unstimulated animals *or* animals with more extensive enriched experience later in life. In other words, the infant experience again appeared to decrease synapse numbers per neuron while facilitating behavioral change. This result is counter intuitive and implies that there are other critical neuronal or glial changes, which we did not measure, that are supporting behavioral change. For instance, it is possible that there are more neurons (or glia) in the animals stimulated early in life. If a brain has more neurons it is quite likely that it has more, not less, total synapses, even though there may be fewer per neuron. This could account for our behavioral results. Furthermore, if the neuron increase were age dependent, it would account for the age-dependent dendritic changes. That is, in the absence of extra neurons, the brain alters the ones that are available, which could lead to greater dendritic length (Fig. 5).

The possibility that neuron number could explain our dendrite-behavior paradox is supported by a recent study by Kempermann et al. (1997). They showed that juvenile mice placed in enriched environments had significantly more neurons in the dentate gyrus of the hippocampus than matched controls. (Although these authors did not study the cortex we can presume that there could be neuron differences in other brain regions.). There are two ways that a brain could have more neurons: either fewer neurons died or new neurons were generated in response to the experience. Kempermann et al. concluded in their study that fewer cells had died. Nonetheless, this does not preclude the possibility that new neurons can be generated in other cortical regions. This would be especially true if experience stimulated activity in the stem cells of the subventricular zone. This is an intriguing possibility that warrants further study.

I have focused upon differences in brain plasticity that are seen in early stages of life and have not considered the effects of aging. The aging brain is capable of plastic changes both in neurons and in glia (e. g., Black et al. 1987). In fact, we have found in our studies of rats in enriched environments that the

A. Neuron number is equal in different conditions

Assume 1000 cortical neurons with 10,000 synapses per neuron in control

Total synapses = 10^7

Enrichment increases synapses per neuron to 11,000

Total synapses = 1.1×10^7 (or 1,000,000 more synapses)

B. Neuron number is higher with enrichment

Assume 1000 cortical neurons with 10,000 synapses per neuron in control

Total synapses = 10^7

Enrichment reduces neuron death or increases neurogenesis with total neurons now at 1100. With 10,000 synapses per neuron,

Total synapses = 1.1×10^7 (or 1,000,000 more synapses)

C. Neuron number and synapse number are higher with enrichment

Assume 1000 cortical neurons with 10,000 synapses per neuron in control

Total synapses = 10^7

Enrichment increases neuron number to 1100 and synapse number to 11,000

Total synapses = 1.21×10^7 (or 2,100,000 more synapses)

Fig. 5. Schematic illustration of the effects of cell number and dendritic branching on cortical morphology. **A.** Experience changes only dendritic morphology and not cell number. Assume that a given cortical area has 1000 neurons. If each neuron has 10,000 connections, then there are 10^7 synapses. If the number of synapses per neuron is increased to 11,000 by experience, then the total is 1.1×10^7, which is one million more synapses than in the control. **B.** Experience changes only neuron number and not dendritic morphology. Assume that the effect of experience increases to the number of neurons to 1100. In this case the total number of synapses is increased to 1.1×10^7 as it was in **A**, but in this case measurements of dendritic morphology would show no change from the experience. Nonetheless, the number of synapses has increased. One implication of **B** is that later experience could now change the morphology of the 1100 neurons, leading to a further increase in synapse numbers, which is illustrated in **C**

changes in spine density are actually greater in older animals than in middle-aged ones. In addition, we found that gonadal hormones continue to influence the aging brain, and their sudden removal, such as in ovariectomy in middle age, produces marked effects on cortical morphology (Stewart and Kolb 1995). It is not known, however, how the manipulation of gonadal hormones in the aging animal might influence its response to environmental manipulations. Finally, it appears that the speed of plastic changes is reduced in the aging brain. One intriguing question concerns the relationship between stem cell activity in the subventricular zone and aging. We might predict that stem cells become less active as they age, although this has not been studied to date.

In sum, there are qualitative differences in cortical plasticity at different ages. The extent of these differences begins to become more apparent as we subject the brain to another experience, namely injury.

What Plastic Change Follows Cortical Injury?

When the brain is injured there are three obvious routes to repair: 1) reorganization of existing circuits; 2) the production of chemical messengers to stimulate the reorganization of existing circuits; and 3) the generation of new neurons, and thus new circuits. In fact, the mammalian brain is capable of all three. Furthermore, the routes of repair are age-dependent. Thus, our studies over the past decade have shown that four distinctly different types of plastic changes occur at different ages (Table 1). First, if the cortex is injured during neurogenesis, which is *in utero* in the rat, there is virtually complete functional recovery that is associated with peculiar cortical morphology. Second, if the cortex is injured from P1–P6, there is a miserable functional outcome that is associated with widespread neuronal atrophy. Third, if the cortex is injured from P7–P12, there is good functional recovery, which is correlated with widespread dendritic growth in cortical pyramidal neurons. In addition, if the damage is restricted to the midline frontal cortex, there is a reintroduction of neurogenesis and the brain regrows (Fig. 3). This is correlated with functional recovery. Fourth, if the cortex is injured in adulthood, there is an initial atrophy of dendritic fields of neurons adjacent to the lesion followed, in some cases, by a gradual regrowth of dendrites, with an eventual net increase. If there is no regrowth there is no functional recovery; if there is regrowth there is partial recovery.

Table 1. Summary of the effects of frontal cortical injury at different ages

Age at injury	Result	Basic reference
E18	Cortex regrows with odd structure Functional recovery	Kolb et al. 1998a
P1–P6	Small brain, dendritic atrophy Dismal functional outcome	Kolb and Gibb 1990
P7–P12	Dendrite and spine growth Cortical regrowth Functional recovery	Kolb and Gibb 1990 Kolb et al. 1998b
P120	Dendritic atrophy, then growth Partial return of function	Kolb 1995
P120+NGF	Dendritic and spine growth Enhanced functional recovery	Kolb et al. 1997

Abbreviations: E18, embryonic day 18; Px, postnatal day x; NGF, nerve growth factor

The age-related changes in recovery correlate with the plastic periods discussed above (Fig. 1). In particular, rats with damage from P1–P6 show the least plasticity, even in comparison to middle-aged animals, whereas rats with damage from P7–P12 show the most plasticity, which includes the spontaneous generation of new neurons. Our discovery that the brain could generate new neurons after an injury is worth more detailed consideration. We have found that both the olfactory bulb and the midline frontal cortex regenerate after injury around P10 but not before P6 or after P15 (Fig. 6 and 7). Furthermore, injury to other cortical regions is not associated with regeneration. Thus, there is something privileged about the anterior part of the telencephalon. There is some difference in the mechanism supporting regeneration of the olfactory bulb and cortex, however. We have been able to completely block regeneration of the cortex in animals treated with bromodeoxyuridine (BrdU), which is a commonly used mitotic marker, if we inject pregnant rats on E13–E15. Earlier or later injections are without effect. In contrast, the BrdU treatment does not affect the regeneration of the olfactory bulb.

In sum, we have shown that brain plasticity after cortical injury is age-dependent and that the brain is capable of both reorganizing existing circuits and generating new ones. The next question is to ask if we can control these processes.

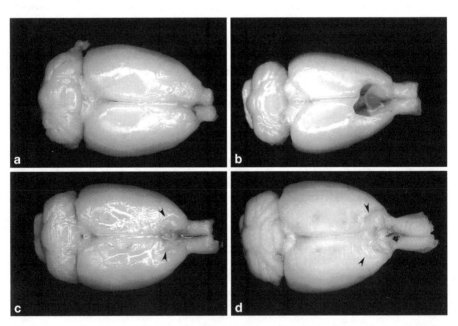

Fig. 6. Photographs of brains of rats that had frontal lesions and various treatments. **A.** Regrowth after frontal lesion on postnatal day 10. The brain looks essentially normal. **B.** Blockade of regrowth in rat treated as in **A**, except that it was given BrdU on embryonic day 13. This illustrates the size of the lesion in the absence of growth. **C and D.** Neurotrophin-induced regrowth after adult frontal lesion. The arrows indicate the border of the original lesion

Fig. 7. Photographs of the brains of rats that had the olfactory bulb removed either on postnatal day 1 (**A**) or day 10 (**B**). The olfactory bulb regrew after the removal on day 10 but not on day 1. In contrast, after removal at day 1 the frontal pole became distorted and moved into the vacant bulb cavity

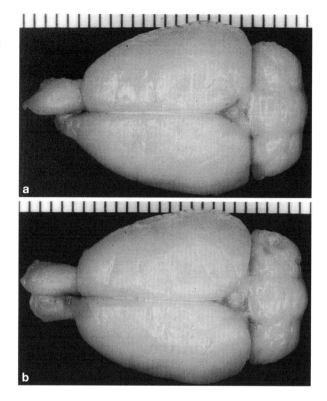

Can Brain Plasticity be Increased?

The ecological theory postulates that it is possible to partially restore lost plasticity by recreating the optimal conditions that were present during development. This process of recreation is likely to require chemical stimulation, such as with compounds like neurotrophins (NTs), or it may require behavioral manipulation, such as in behavioral therapies. Furthermore, there is good reason to believe that behavioral manipulation may influence the endogenous production of neurotrophins, so the two may be related.

Increasing Plasticity in the "Normal" Brain

We have seen that the normal brain is responsive to various types of environmental manipulation ranging from tactile stimulation in infants to enriched housing or the learning of specific motor or cognitive tasks (for reviews, see Kolb et al. 1998; Kolb and Whishaw 1998). It has been shown that environmental experience may increase the production of NTs by the brain (e.g., Schoups et al. 1995), which implies that the NTs may play a role in the observed plasticity. If so, we can predict that 1) adding NTs might stimulate dendritic or other change, and 2) adding NTs to animals given specific experience might enhance the plastic changes.

In one study we infused nerve growth factor (NGF) into the lateral ventricle of rats for seven days (Kolb et al. 1997b). Two months later we examined the dendritic arborization and found a dramatic increase in dendritic arborization and in spine density (Fig. 8). These changes were reminiscent of those seen in rats housed as adults in enriched environments, although even more extensive. We have preliminary results suggesting that epidermal growth factor (EGF) and fibroblast growth factor (bFGF) have similar actions, although we do not yet know if the actions of the three NTs are identical. We also have preliminary evidence that a cocktail of NGF, EGF, and bFGF results in a marked increase in the activity of the stem cells in the subventricular zone of adult rats. These are the cells that produce cortical neurons and astrocytes, which implies that the NTs could induce the adult brain to produce new neurons and/or astrocytes. We return to this below.

Another approach to increasing plasticity is to stimulate the production of other chemicals in the brain. Because it has long been known that enriched rearing leads to increased acetylcholine (ACh) production, it seemed reasonable to

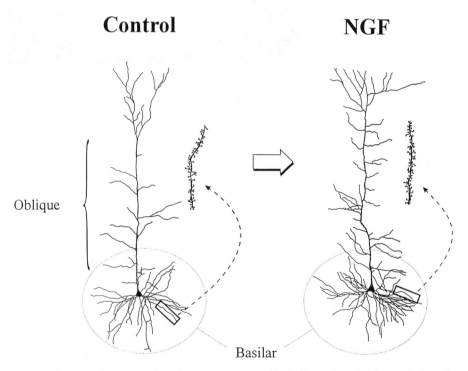

Control   **NGF**

Oblique

Basilar

Fig. 8. Schematic illustration of the effects of infusion of nerve growth factor (NGF) into the lateral ventricle of adult rats. Both the dendritic branching and the spine density were increased in neurons throughout the cerebral cortex (after Kolb et al. 1997)

predict that increased ACh might be related to increased plasticity. To test this idea we fed pregnant (and later lactating) rats a diet rich in choline (R. Tees and B. Kolb, unpublished research). The pups showed increased ACh levels in the cortex and had increased dendritic complexity. Again, it appears that chemical intervention can influence cortical structure. The critical question now is to determine to what extent NTs or choline (or other chemicals) can increase the brain's response to sensory experience and to what extent this is manifest in potentiated behavioral capacities.

Increasing Plasticity in the Injured Brain

We have seen that the injured brain shows a plastic response that varies with age. We can make two predictions from the theory. First, we ought to be able to induce plastic changes, and subsequent recovery, in animals that show poor recovery to cortical injury. This could be done with manipulations of sensory experience or by adding NTs or chemicals such as choline. Second, animals that show maximal plasticity and recovery after cortical injury should show very little subsequent change in response to our interventions. Both hypotheses are confirmed.

Our first studies took advantage of the developmental differences in plasticity (Fig. 1). We reasoned that if a brain were injured at birth, we should be able to at least partially reverse the dismal functional outcome by stimulating the brain when it is maximally responsive, namely from about P10–P30. We therefore gave rats frontal lesions on P4 or P10 and then gave some animals tactile stimulation until weaning (Fig. 9). In adulthood the animals were tested on a variety of behavioral tasks. The results were clear. First, animals with P4 lesions showed dramatic recovery if they received tactile stimulation. Animals with P10 lesions showed no additional recovery with this treatment (Fig. 9). Second, animals with P4 lesions showed significant functional recovery with enrichment either as juveniles or in adulthood, although the recovery was less impressive than with the tactile stimulation. Again, rats with P10 lesions showed little benefit from this treatment.

A second series of experiments gave NTs to rats with lesions in adulthood. For example, in one experiment we infused NGF into the ventricles of rats with large unilateral strokes (Kolb et al. 1997). Two months later we assessed behavioral recovery and later we analyzed dendritic changes. The NGF treatment enhanced recovery and this was correlated with dendritic growth. A later study showed parallel results with bFGF (Rowntree and Kolb 1997).

A third series of experiments tried a different approach. In this case we reasoned that because we had shown that the infant brain was capable of regeneration, we ought to be able to induce regeneration in adult rats with an appropriate cocktail of NTs. We now have shown that a combination of EGF, NGF, and bFGF will induce the brain to generate new cells and that these cells will migrate to the lesion site, and later they will differentiate in neurons and glia (Fig. 10). Many,

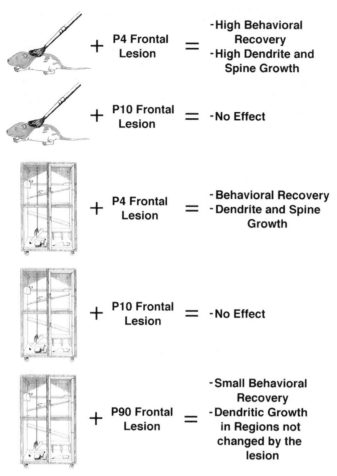

Fig. 9. Schematic summary of the effects of sensory experience on recovery from cortical injury at different ages. Note that the brains of animals that show little spontaneous dendritic changes after cortical injury (that is, at P4) show extensive dendritic (and behavioral) response to sensory experience whereas the brains of animals that show a significant remodeling in response to injury (that is, at P10) show no response to the sensory experience. Furthermore, rats with lesions at P120 show a restricted remodeling of cortical neurons after injury (i.e., only in cortex adjacent to the lesion) and these do not change further with experience. In contrast, the more posterior neurons that were unaffected by the injury do change with experience. These results imply that there are limits to the amount of plasticity that cells can exhibit

although not all, animals with frontal lesions show functional recovery. The critical question is whether the regenerated tissue functions to support functional recovery. At this point we must be cautious in our interpretation. We already know that infusion of the NTs alters the remaining brain, so functional recovery could be related *either* to the new tissue or to the changes in the rest of the brain. This awaits further study.

Do Drugs Produce Enduring Neural Changes?

The repeated intermittent administration of many drugs of abuse results in a progressive increase in their psychomotor activating and rewarding effects, a phenomenon known as behavioral sensitization (e.g., Robinson and Becker 1986). This is an example of experience-dependent behavioral change that is long-lasting, as animals remain hypersensitive for months to years. It is an especially intriguing phenomenon because the neuroadaptations that underlie behavioral sensitization may contribute to drug-induced psychopathology in humans (e.g., Segal and Schuckit 1983; Robinson and Berridge 1993). The Ecological Theory predicts that this experience should produce plastic changes similar to those resulting from sensory experience or injury. In collaboration with Terry Robinson at the University of Michigan, we have shown (Robinson and Kolb 1997) that rats sensitized to amphetamine have significant dendritic growth and an increase in spine density in prefrontal cortex (and nucleus accumbens) but not in other cortical areas (Fig. 10). We can conclude that, like sensory experience, neurotrophins, and brain injury, additive drugs produce long-lasting changes in cortical circuitry. I should note here that our study of the morphological effect of amphetamine is the first in which we have also examined a subcortical structure (nucleus accumbens), and we found large experience-dependent changes. There have been reports of similar changes in the striatum of animals exposed to enriched environments (e.g., Comery et al. 1996). This suggests that the ecological theory of cortical organization may have applications that are far broader than just the cerebral cortex!

One interesting prediction of the similarity between drug-induced and other forms of plasticity is that addictive drugs might influence the plastic changes associated with other types of experience-dependent change. For example, amphetamine might influence the neural changes underlying recovery from brain injury. In fact, amphetamine has been claimed to enhance recovery, at least in some circumstances (e.g., Feeney and Sutton 1987), although the mechanism is unknown. We might also predict that drug-induced changes might interact with the effects of sensory experience. Indeed, it is known that drug sensitization is environment-dependent. Again, the mechanism of this is unknown.

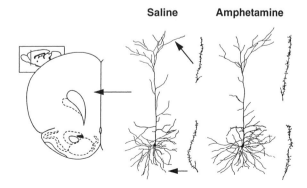

Fig. 10. Camera lucida drawings of representative layer III pyramidal cells in the prefrontal cortex (area Cg3) of saline- and amphetamine-pretreated rats. The drawing to the right of each cell represents an apical or basilar dendritic segment used to calculate spine density (after Robinson and Kolb 1997).

Conclusions

I have proposed that one way to understand the nature of brain plasticity is to consider an ecology of the cerebral cortex (for more extensive reviews, see Kolb et al. 1998). On the basis of my research and that of my colleagues, as well as that of others, I suggest that an ecological theory of the cortical organization would have the following 10 postulates.

1. The cortex changes throughout life. This capacity is known as *plasticity.*
2. Cortical plasticity functions to reorganize the cortical structure that is defined by a basic genetic template. Such a template cannot specificy all possible connections, however, and this leaves considerable flexibility in the ultimate organization of the cortex. The ultimate effect of the various plastic changes is either to modify existing cortical circuits or to create novel circuitry.
3. Cortical plasticity varies with age. Maximum plasticity will occur at that time when the cortex has the most mechanisms at its disposal. This occurs at a critical period during development. Minimal plasticity will occur when the cortex has the least mechanisms at its disposal. This probably occurs during senescence.
4. When the cortex changes, this is reflected in behavioral change. Behavioral change may be referred to by many names, including learning, maturation, recovery, addiction, etc. For example, environmental stimulation produces increased dendritic growth and, in adults, increased spine density in cortical pyramidal cells. Animals show enhanced behavioral capacities after such experience, and cortical reorganization. Similarly, when the cortex is damaged at certain times during life, there is a subsequent compensatory growth of dendrites and increased spine density that is correlated with behavioral recovery. At other times in life there is no compensatory change and no functional recovery.
5. The environment of the cortex is altered by experience. Experience includes the effect of sensory events, gonadal hormones, aging, brain injury, stress, and drugs and other chemicals. The effects of experience on the cortex are age-dependent.
6. There are multiple mechanisms of plasticity that range from gross cortical changes, such as the generation of neurons and glia, to more subtle changes such as the alteration of synapses or changes in the production of chemical messengers.
7. It is possible to partially restore lost plasticity by recreating the developmental conditions that favor maximal plasticity. This may involve chemical stimulation, such as with neurotrophins (NTs), or behavioral manipulation, such as in behavioral therapies.
8. The induction of plastic change may involve multiple steps. Thus, one reason that behavioral therapies may be effective in changing the cortex is that the behavioral changes brought about by specific experiences actually stimulate

the brain to produce neurotrophins. The neurotrophins then act to stimulate change in the cortex.

9. The mechanisms underlying the plasticity resulting from different experiences are similar, although not necessarily identical. Thus, the neural or glial changes observed in response to sensory experience, brain injury, and addictive drugs are remarkably similar. The conclusion from this is that there may be basic mechanisms of synaptic change in the mammalian cortex that are used in many forms of plasticity. This implies that the effects of different experiences may interact. For example, we would predict that the effect of addictive drugs on the brain might vary with past experiences such as stress, brain injury, or specific sensory events.

10. When the cortex changes, its capacity for subsequent change is compromised. For example, when the cortex is damaged, it changes to compensate for the injury and is less able to change in response to other experiences. This implies that the most potent time for behavioral therapies might be while the brain is spontaneously reorganizing because the experience ought to influence the reorganization. Another example is seen in the effect of gonadal hormones. Once the cortex is exposed to gonadal hormones, it is altered. As a result, the effect of subsequent experiences is changed accordingly.

Finally, one prediction that follows from postulate 10 is that there are likely to be limits to the amount that a brain can change. We still do not known what determines the limits or, in most cases, what the limits might be. Nonetheless, the capacity of the cortex to change is constrained by the life history of the individual or, stated differently, by the ecology that brain finds itself in.

Acknowledgements

This research was supported by grants from Canadian NSERC, MRC, and NCE. I wish to thank Robbin Gibb, Grazyna Gorny, Reed Kindt, Terry Robinson, and Ian Q. Whishaw for their help both in the execution of the experiments and the development of the theoretical model.

References

Black JE, Greenough WT, Anderson BJ, Isaacs KR (1987) Environment and the aging brain. Can J Psychol 41:111–130

Comery TA, Stamoudis, CX, Irwin SA, Greenough WT (1996) Increased density of multiple-head dendritic spines on medium-sized spiny neurons of the striatum in rats reared in a complex environment. Neurobiol Learn Mem 66:93–96

Feeney DM, Sutton RL (1987) Pharmacotherapy for recovery of function after brain injury. CRC Crit Rev Neurobiol 3:135–197

Greenough WT, Chang F-LF (1985) Synaptic structural correlates of information storage in mammalian nervous systems. In: Cotman CW (ed) Synaptic plasticity. Guilford, New York, pp 335–372

Hebb DO (1949) The organization of behavior. McGraw-Hill, New York

Kempermann G, Kuhn HG, Gage FH (1997) More hippocampal neurons in adult mice living in an enriched environment. Nature 386:493–495

Kolb B (1995) Brain plasticity and behavior. Lawrence Erlbaum, Mahwah, NJ

Kolb B, Gibb R (1990) Anatomical correlates of behavioural change after neonatal prefrontal lesions in rats. Prog Brain Res 85:241–256

Kolb B, Whishaw IQ (1988) Brain plasticity and behavior. Ann Rev Psychol, in press

Kolb B, Cioe J, Muirhead D (1998a) Cerebral morphology and functional sparing after prenatal frontal cortex lesions in rats. Behav Brain Res 91:143–155

Kolb B, Cote S, Ribeiro-da-Silva, Cuello AC (1997b) NGF stimulates recovery of function and dendritic growth after unilateral motor cortex lesions in rats. Neuroscience 76:1139–1151

Kolb B, Forgie M, Gibb R, Gorny G, Rowntree S (1998) Age, experience, and the changing brain. Neurosci Biobehav Rev 22:143–159

Kolb B, Gibb R, Gorny G, Whishaw IQ (1997c) Possible brain regrowth after cortical lesions in rats. Behav Brain Res, in press

Kolb B, Gorny G, Cote S, Ribeiro-da-Silva, Cuello AC (1997d) Nerve growth factor stimulates growth of cortical pyramidal neurons in young adult rats. Brain Res 751:289–294

Konorski J (1948) Conditioned reflexes and neuron organization. Cambridge University Press, Cambridge

Ramon y Cajal S (1928) Degeneration and regeneration of the nervous system. Oxford, London

Robinson TE, Becker JB (1986) Enduring changes in brain and behavior produced by chronic amphetamine administration: a review and evaluation of animal models of amphetamine psychosis. Brain Res Rev 11:157–198

Robinson TE, Berridge KC (1993) The neural basis of drug craving: an incentive-sensitization theory of addiction. Brain Res Rev 18:247–291

Robinson TE, Kolb B (1997) Persistent structural adaptations in nucleus accumbens and prefrontal cortex neurons produced by prior experience with amphetamine. J Neurosci 17:8491–8498

Rowntree S, Kolb B (1997) Antibodies to bFGF block functional recovery and dendritic compensation after motor cortex lesions. Eur J Neurosci 9:2432–2442

Schoups AA, Elliott RC, Friedman WJ, Black IB (1995) NGF and BDNF are differentially modulated by visual experience in the developing geniculocortical pathway. Dev Brain Res 86:326–334

Segal DS, Schenkit MA (1983) Animal models of stimulant-induced psychosis. In: Creese I (ed) Stimulants: neurochemical, behavioral and clinical perspectives. Raven, New York, pp 131–167

Stewart J, Kolb B (1994) Dendritic branching in cortical pyramidal cells in response to ovariectomy in adult female rats: suppression by neonatal exposure to testosterone. Brain Res 654:149–154

Uylings HBM, van Eden CG, Parnavelas JG, Kalsbeek A (1990) The prenatal and postnatal development of rat cerebral cortex. In: Kolb B, Tees R (eds) The cerebral cortex of the rat. MIT Press, Cambridge, MA, pp 35–76

Pain and Neuroplasticity

R. Melzack, T. J. Coderre[1], *A. L. Vaccarino*[2] *and J. Katz*[3*]

Summary

The traditional specificity theory of pain perception holds that pain involves a direct transmission system from somatic receptors to the brain. The amount of pain perceived, moreover, is assumed to be directly proportional to the extent of injury. Recent research, however, indicates far more complex mechanisms. Clinical and experimental evidence shows that noxious stimuli may sensitize central neural structures involved in pain perception. Salient clinical examples of these effects include amputees with pains in a phantom limb that are similar or identical to those felt in the limb before it was amputated, and patients after surgery who have benefited from pre-emptive analgesia, which blocks the surgery-induced afferent barrage and/or its central consequences. Experimental evidence of these changes is illustrated by the development of sensitization, wind-up or expansion of receptive fields of CNS neurons, as well as by the enhancement of flexion reflexes and the persistence of pain or hyperalgesia after inputs from injured tissues are blocked. It is clear from the material presented that the perception of pain does not simply involve a moment-to-moment analysis of afferent noxious input, but rather involves a dynamic process that is influenced by the effects of past experiences. Sensory stimuli act on neural systems that have been modified by past inputs, and the behavioural output is significantly influenced by the "memory" of these prior events. An increased understanding of the central changes induced by peripheral injury or noxious stimulation should lead to new and improved clinical treatment for the relief and prevention of pathological pain.

Introduction

Pain research and therapy, at any period in history, are determined by the dominant theory of brain function at the time. Until the last half of this century, pain was thought to be produced by a passive, direct-transmission system from

---

[1] Now at the Institut de Recherche Clinique de Montreal, Montreal, Quebec, Canada.
[2] Now at the Department of Psychology, University of New Orleans, New Orleans, Louisiana, U.S.A.
[3] Now at the Department of Psychology, Toronto Hospital, Toronto, Ontario, Canada.
[*] Department of Psychology, McGill University, 1205 Dr. Penfield Avenue Montreal, Quebec, H3A 1B1

J. Grafman / Y. Christen (Eds.)
Neuronal Plasticity:
Building a Bridge from the Laboratory to the Clinic
© Springer-Verlag Berlin Heidelberg New York 1999

peripheral receptors to cortex. There was no place in this concept of the nervous system for "plasticity," in which neuronal and synaptic functions are capable of being molded or shaped so that they influence subsequent perceptual experiences. Plasticity related to pain represents persistent functional changes, or "somatic memories" (Katz and Melzack 1990), produced in the nervous system by injuries or other pathological events. The recognition that such changes can occur is essential to understanding the chronic pain syndromes, such as low back pain and phantom limb pain, that persist and often destroy the lives of the people who suffer them.

The theory of pain that we inherited in the 20th century was proposed by Descartes three centuries earlier (see Melzack and Wall 1996). It holds that injury activates specific pain receptors and fibers which, in turn, project pain impulses through a spinal pain pathway to a pain center in the brain. The psychological experience of pain, therefore, was virtually equated with physical injury. In the 1950s, there was no room for psychological contributions to pain, such as attention, past experience and the meaning of the situation. Instead, pain experience was held to be proportional to peripheral injury or pathology. Patients who suffered chronic pain without presenting signs of organic disease were often sent to psychiatrists.

In 1965, Melzack and Wall proposed the gate control theory of pain. The theory's emphasis on the modulation of inputs in the spinal dorsal horns and the dynamic role of the brain in pain processes had a clinical as well as a scientific impact. Psychological factors, which were previously dismissed as "reactions to pain," were now seen to be an integral part of pain processing and new avenues for pain control were opened. Similarly, cutting nerves and pathways was gradually replaced by a host of methods to modulate the input. Physical therapists and other health care professionals who use a multitude of modulation techniques (including acupuncture) were brought into the picture, and transcutaneous electrical nerve stimulation (TENS) became an important modality for the treatment of chronic and acute pain (Melzack and Wall 1996).

The gate control theory's most important contribution to biological and medical science was its emphasis on central nervous system (CNS) mechanisms. The theory forced the medical and biological sciences to accept the brain as an active system that filters, selects and modulates inputs. The dorsal horns, too, were not merely passive transmission stations but sites at which dynamic activities – inhibition, excitation and modulation – occurred. The theory highlighted the CNS as an essential component in pain processes.

Even though the Cartesian concept of direct transmission has dominated our ideas about pain for the past 300 years, descriptions of plasticity related to pain – that is, the idea that injury can produce alterations in CNS function affecting subsequent pain sensitivity – have been proposed by a few courageous clinical observers. MacKenzie (1893) suggested that increased pain sensitivity and referred pain could be the result of increased sensitivity of CNS structures. He proposed that sensory impulses arising from injured tissues create an "irritable focus" in spinal cord segments onto which they impinge. In relation to perioperative anesthesia, Crile (1913) wrote that patients given inhalational anesthesia still need to be protected by regional anesthesia; otherwise they might incur persistent CNS changes and enhanced post-operative pain. According to Hardy et al. (1950), secondary hyperalgesia and referred cutaneous hyperalgesia occur because an injury produces a state of hyperexcitability in the spinal cord. This hyperexcitability is sustained following the activation of a network of internuncial neurons, which produces a spreading facilitation of adjacent neurons in the spinal cord, allowing for the spread of hyperalgesia to uninjured regions of the body. Similarly, Livingston (1943) suggested that the afferent activity generated by injured peripheral nerves elicits an abnormal firing pattern within the spinal cord. He proposed that a disturbance occurs in an internuncial pool of dorsal horn interneurons and results in reverberatory activity that eventually spreads to other parts of the spinal cord, including areas that affect the sympathetic chain. Increased activity in sympathetic efferents would disrupt vasoregulation and induce further hypersensitivity of peripheral tissue, leading to increased afferent input and a "vicious circle" of peripheral-central activity.

Aside from descriptive references to irritable foci, reverberatory activity and vicious circles, the above theories do not provide empirical evidence for, or details of, the nature of the CNS changes that occur following noxious stimulation. Only recently has there been specific empirical evidence indicating noxious stimulus-induced changes in CNS function. Kenshalo et al. (1982) demonstrated that noxious peripheral stimuli produce changes in the sensitivity of dorsal horn neurons to further stimulation, and Woolf and Wall (Woolf 1983; Woolf and Wall 1986a) provided empirical evidence for a primary afferent input triggering sustained increases in central excitability. Woolf (1983) demonstrated that injury-induced increases in spinal cord excitability could be maintained even after local anesthesia of the injured site, providing empirical evidence that acute injury could produce lasting spinal changes. Woolf and Wall (1986a) showed that the amount of morphine required to prevent the development of this spinal hyperexcitability was 10-fold less than the amount required to reverse it after it was established, and provided the experimental basis for subsequent clinical investigations of the use of pre-emptive analgesia for the prevention or alleviation of post-operative pain.

These studies indicate that noxious stimulation or injury can produce dramatic alterations in spinal cord function, including sensitization, wind-up or the expansion of the receptive fields of spinal neurons. Recently, several investigators have proposed detailed theories of how noxious stimuli produce these changes in CNS function. Unlike previous theories of central sensitization, recent theories propose that, in addition to a contribution of neuronal hyperactivity to pathological pain, there are specific cellular and molecular changes that affect membrane excitability and induce new gene expression, thereby allowing for enhanced responses to future stimulation. These studies have recently been reviewed by Coderre et al. (1993, 1997). The effects of these changes include an expansion of dorsal horn receptive fields and hyperexcitability which, if allowed to persist, would presumably produce prolonged changes in excitability that could be maintained without further noxious peripheral input.

Pain in Phantom Limbs and Deafferented Structures

A striking property of phantom limb pain is the persistence of a pain that existed in a limb prior to its amputation. This type of phantom limb pain, characterized by the persistence or recurrence of a previous pain, has the same qualities and is experienced in the same area of the limb as the pre-amputation pain. Case studies of amputees (see Katz and Melzack 1990) have demonstrated pain "memories" of painful diabetic and decubitus ulcers, gangrene, corns, blisters, ingrown toe-nails, cuts and deep tissue injury. In addition, the phantom limb may assume the same painful posture as that of the real limb prior to amputation, especially if the arm or leg had been immobilized for a prolonged period (Katz and Melzack 1990).

The literature indicates that the proportion of amputees who report that their phantom pains are similar to those felt in the limb before amputation may be as high as 79 % (Katz and Melzack 1990). Reports of pain memories in phantom limbs appear to be less common when there has been a discontinuity, or a pain-free interval, between the experience of pain and the amputation. This is consistent with the observation that relief of pre-amputation pain by continuous epidural block for three days prior to amputation decreases the incidence of phantom limb pain six months later (Bach et al. 1988). Furthermore, if pain is experienced at or near the time of amputation, there is a higher probability that it will persist in the phantom limb (Jensen et al. 1985; Katz and Melzack 1990).

Pain also persists in patients with deafferentation that does not involve amputation. Patients with brachial plexus avulsions (Jensen and Rasmussen 1989) or spinal cord injuries (Berger and Gerstenbrand 1981) often experience pain in the anesthetic, deafferented region. For example, Nathan (1962) described a patient who continued to feel the pain of an ingrown toe-nail after a complete spinal cord break. In addition, patients undergoing spinal anesthesia (Wallgren 1954) and those with injuries of the brachial plexus (Jensen and Rasmussen 1989) or spinal cord (Berger and Gerstenbrand 1981) sometimes report that a limb is in the same uncomfortable, often painful posture it was in prior to the injury or block. These postural phantom sensations do not usually persist beyond several days, and in most cases are at least temporarily reversed by competing visual inputs that reveal a dissociation between the real and perceived limb.

There is also literature on the persistence of painful and non-painful sensations associated with removal or deafferentation of body structures other than the limbs, including breasts (Kroner et al. 1989), teeth (Hutchins and Reynolds 1947; Reynolds and Hutchins 1948), and internal and special sense organs. Ulcer pain has been reported to persist after vagotomy (Szasz 1949) or subtotal gastrectomy with removal of the ulcer (Gloyne 1954). Similarly, patients have reported labor pain and menstrual cramps following total hysterectomy (Dorpat 1971), rectal and hemorrhoid pain following removal of the rectum (Ovesen et al. 1991), the burning pain of cystitis after complete removal of the bladder (Brena and Sammons 1979), and the pain of a severely ulcerated cornea after enucleation of an eye (Minski 1943).

When a missing or completely anesthetic limb continues to be the source of pain that resembles an old injury, it is reasonable to assume that the pain is centrally represented, but it is not clear whether deafferentation per se is necessary for pain memories to develop. The interruption of afferent input associated with deafferentation may facilitate the central neural changes that contribute to the formation of pain memories by removing normal inhibitory control mechanisms. In addition, since amputation also results in the loss of visual and tactile information related to the limb, the central influences that normally inhibit the established pain "traces" may be reduced further by the absence of information from external sources that could confirm or disconfirm the percept arising from the peripheral injury.

There is evidence that in some instances the reactivation of pain memories requires a peripheral trigger. Leriche (1947) described a patient who did not experience phantom limb pain until six years after amputation, when an injection into the stump instantly, and permanently, revived the pain of a former painful ulceration of the Achilles tendon. Nathan (1962, 1985) reported a similar phenomenon when applying noxious stimulation to the stump of an amputee who later re-experienced the pain of an ice-skating injury he had sustained five years earlier when the leg was intact. Noordenbos and Wall (1981) described seven patients with partial peripheral nerve injury, and subsequent pain, who underwent complete nerve resection and graft or ligation. Following regeneration and a pain-free period, all re-developed pain of the same quality and in the same location as the pain they had experienced prior to nerve resection, although in some patients the recurrence of pain was restricted to a smaller area within the originally painful region. These studies and case reports indicate that past pains may be reactivated months or even years after the original injury, in some case by a peripheral trigger that provides that input required to activate the central neural structures subserving the memory trace.

Deafferentation by peripheral neurectomy or dorsal rhizotomy in rodents is followed by self-mutilation (autotomy) in which the animals bite and scratch the insensate paw to the point of amputation (Wall et al. 1979). There is evidence that autotomy behavior is produced by ongoing pain or dysesthesia, associated with increased neuronal activity, which is referred to the anesthetic region (Blumenkopf and Lipman 1991). Autotomy behavior is dramatically affected by alterations in the level of noxious input present at the time of, or prior to, nerve section. Thus, noxious chemical (Dennis and Melzack 1979; Coderre et al. 1986), thermal (Coderre and Melzack 1987; Katz et al. 1991) and electrical (Katz et al. 1991; Seltzer et al. 1991) stimulation prior to nerve sections significantly increases the severity of autotomy following neurectomy or rhizotomy. These findings suggest that the prior injury produces central changes that influence nociceptive behavior after nerve sections, at a time when inputs from the injured region are no longer capable of transmitting their message centrally.

The above findings are similar to clinical reports that phantom limb pain is more likely to occur in amputees who had pain in their limb prior to amputation, and strongly suggest that central neuroplasticity is crucial to the development of phantom limb pain. The clinical relevance of these findings is indicated by the observation that in human amputees the incidence of phantom limb pain at seven days and six months after amputation is significantly greater in patients whose pain is not treated by epidural block with bupivacaine and morphine prior to amputation surgery (Bach et al. 1988). In contrast to the effect of increasing noxious inputs at the time of nerve injury, reducing or eliminating the afferent barrage induced by nerve section produces a dramatic reduction in autotomy. When the afferent barrage induced by nerve cuts in rats is blocked by treating the sciatic and saphenous nerves with local anesthetics prior to sectioning them, there is a significant reduction in the incidence and severity of autotomy (Seltzer et al. 1991).

An animal model has recently been developed (Katz et al. 1991) which parallels the observation that human amputees report similar pains in a limb before and after amputation. In this animal model, rats selectively initiated autotomy in either the lateral or medial half of a hindpaw if that particular half had been given a thermal injury prior to sciatic and saphenous nerve sections. The selective attack on the previously injured region, despite the fact that the entire foot was deafferented, suggests that the rats were responding to pain referred to the injured area, which was produced by the prior injury and the central trace it created. Rats injured after neurectomy did not show a similar preference, indicating that the rats were not responding simply to peripheral cues associated with the injury.

Denervation Hypersensitivity and Neuronal Hyperactivity

Sensory disturbances associated with nerve injury have been closely linked to alterations in CNS function. Markus et al. (1984) demonstrated that the development of hypersensitivity in a rat's hindpaw following sciatic nerve section occurs concurrently with the expansion of the saphenous nerve's somatotopic projection in the spinal cord. Nerve injury may also lead to the development of increased neuronal activity at various levels of the somatosensory system. In addition to spontaneous activity generated from the neuroma (Wall and Gutnik 1974), peripheral neurectomy also leads to increased spontaneous activity in the dorsal root ganglion (Wall and Devor 1983), dorsal spinal roots (Howe et al. 1977) and spinal cord (Asada et al. 1990). Furthermore, after dorsal rhizotomy, there are increases in spontaneous neural activity in the dorsal horn (Basbaum and Wall 1976), the spinal trigeminal nucleus (Macon 1979) and the thalamus (Albe-Fessard and Lombard 1983).

Clinical neurosurgery studies reveal a similar relationship between denervation and CNS hyperactivity. Neurons in the somatosensory thalamus of patients with neuropathic pain display high spontaneous firing rates, abnormal bursting activity, and evoked responses to stimulation of body areas that normally do not activate these neurons (Lenz et al. 1989; Rinaldi et al. 1991). The site of abnormality in thalamic function appears to be somatotopically related to the painful region. In patients with complete spinal cord transection and dysesthesias referred below the level of the break, neuronal hyperactivity was observed in thalamic regions that had lost their normal sensory input, but not in regions with apparently normal afferent input (Lenz et al. 1987). Furthermore, in patients with neuropathic pain, electrical stimulation of subthalamic, thalamic and capsular regions may evoke pain and in some instances even reproduce the patient's pain (Nathan 1985; Tasker 1989). Direct electrical stimulation of spontaneously hyperactive cells evokes pain in some but not all pain patients, raising the possibility that in certain patients the observed changes in neuronal activity may contribute to the perception of pain (Lenz et al. 1987). Studies of patients undergoing electrical brain stimulation during brain surgery reveal that pain is rarely elicited by test stimuli unless the patient suffers from a chronic pain problem. However, brain stimulation can elicit pain responses in patients with chronic pain that does not involve extensive nerve injury or deafferentation. Nathan (1985) describes a patient who underwent thalamic stimulation for a movement disorder. The patient had been suffering from a toothache for 10 days prior to the operation. Electrical stimulation of the thalamus reproduced the toothache.

It is possible that receptive field expansions and spontaneous activity generated in the CNS following peripheral nerve injury are, in part, mediated by alterations in normal inhibitory processes in the dorsal horn. Within four days of a peripheral nerve section there is a reduction in the dorsal root potential, and therefore, in the presynaptic inhibition it represents (Wall and Devor 1981). Nerve section also induces a reduction in the inhibitory effect of A-fiber stimulation on activity in dorsal horn neurons (Woolf and Wall 1982). Furthermore, nerve injury affects descending inhibitory controls from brainstem nuclei. In the intact nervous system, stimulation of the locus coeruleus (Segal and Sandberg 1977) or the nucleus raphe magnus (Oliveras et al. 1979) produces an inhibition of dorsal horn neurons. Following dorsal rhizotomy, however, stimulation of these areas produces excitation, rather than inhibition, in half of the cells studied (Hodge et al. 1983).

Effects of Anesthetic or Analgesic Pretreatment on Post-Injury Pain

As noted above, deafferentation pain in rats is significantly reduced if the injured nerves are locally anesthetized prior to nerve injury. Thus, autotomy after nerve sections (Seltzer et al. 1991), or hyperalgesia following nerve ligation (Dougherty et al. 1992), is significantly reduced if the sciatic and saphenous nerves are locally anesthetized prior to the nerve injury. Recent evidence indicates that persistent pain induced by tissue injury is also reduced by pretreatment with local anesthetics or opioids prior to the injury, suggesting a contribution of central plasticity to nociceptive pain. A subcutaneous injection of dilute formalin produces a biphasic nociceptive response with an early phase of intense pain that occurs in the first few minutes and a later tonic phase of moderate pain occurring about 20–60 min after formalin injection (Dubuisson and Dennis 1977). The nociceptive response to subcutaneous formalin is matched by a corresponding biphasic increase in the activity of dorsal horn neurons after formalin injection (Dickenson and Sullivan 1987a). Dickenson and Sullivan (1987b) demonstrated that intrathecal administration of a mu-opiate agonist significantly inhibits the prolonged increase in dorsal horn activity produced by subcutaneous formalin injection. However, this inhibition occurs only if the drug is given before the formalin injection, and not if it is given two minutes after the injection. These results imply that the dorsal horn activity associated with the late phase of the formalin test depends upon spinal activation during the early phase immediately after formalin injection.

Behavioral studies support the electrophysiological finding that the late phase response to formalin is, in part, dependent on spinal changes generated during the early phase. Tonic nociceptive responses in the late phase of the formalin test (30–60 min after formalin) are not eliminated by complete anesthetic blockade of the formalin injected area at the time of testing during the late phase, but are virtually abolished if the area was also blocked by local anesthetics at the time of formalin injection (Coderre et al. 1990). Furthermore, late phase nociceptive responses are significantly reduced by spinal anesthesia induced immediately prior to formalin injection, but not by spinal anesthesia administered five min after formalin injection, that is, after the early phase had already occurred (Coderre et al. 1990). These results suggest that central neural changes, which occur during the early phase of the formalin test, are essential for the development of the later tonic phase of the formalin test.

Evidence suggests that peripheral tissue injury also induces plasticity in supraspinal structures, which affects persistent pain behavior. This evidence comes from assessing the effects of pre-injury treatment with local anesthetics (in this case injected into discrete brain regions) on post-injury pain responses. Nociceptive responses to subcutaneous formalin injection into the rat hindpaw are suppressed after focal injection of lidocaine into specific limbic system sites such as the cingulum bundle and the fornix pathway. The lidocaine injection produces analgesia during the late phase of the formalin test (30–70 min after formalin injection) when injected into these areas 10 min before, but not 10 min after, the formalin injection (Vaccarino and Melzack 1992). These results suggest that activity in the cingulum bundle and fornix during the early-phase response to formalin is critical to the development of the late-phase response to formalin. The cingulum bundle and fornix are part of a neural loop that projects from the anterior thalamic nuclei to the cingulate cortex, hippocampus and mammillary bodies, and returns to the anterior thalamic nuclei (Vinogradova 1975). It is proposed that activation of this "closed" circuit during the early phase of the formalin response induces a sensitized state within the limbic system, enhancing responses to subsequent stimulation. Recent physiological evidence supports this concept. Brainstem stimulation has been found to enhance the responsiveness of the anterior thalamic nuclei to stimulation of the mammillary bodies and cingulate cortex (Pare et al. 1990). Furthermore, noxious peripheral stimulation produces bursting activity in CA1 neurons of the hippocampus (Sinclair and Lo

1986). The selective blocking of neural activity in the cingulum bundle of fornix during the early phase of formalin may reduce nociceptive responses by preventing the development of long-term changes in these structures.

Post-Operative Pain

The idea that CNS changes by tissue damage and noxious inputs associated with surgery could contribute to post-operative pain has existed for several decades (Crile 1913). However, it was only after the research by Woolf and Wall (1986a) provided a sound justification for pre-emptive treatment that this idea began to receive the clinical attention it deserves. Woolf and Wall (1986a) demonstrated in experimental animals that opioids are much more effective at reducing stimulus-induced increases in the excitability of the dorsal horn if they are administered prior to, rather than following, C-fiber electrical nerve stimulation. Recent clinical evidence supports the hypothesis that the administration of analgesic agents prior to surgery may prevent the central sensitizing effects of the surgical procedure. In this manner it may be possible to reduce post-operative pain intensity or lower post-operative analgesic requirements for periods much longer than the duration of action of the pre-operatively administered agents.

McQuay et al. (1988) examined the possible prophylactic effect of opiate premedication and/or local anesthetic nerve blocks on post-operative pain. They provided data showing that the time to first request for post-operative analgesics was longest among patients who had received a pre-surgical treatment with opiates and nerve blocks, and shortest among patients who had received neither. Similar findings have recently been reported by Kiss and Kilian (1992), who showed that opiate pretreatment increased the length of time until request for first analgesic, reduced the percentage of patients requesting analgesics, and decreased analgesic consumption in the first 48 hours for patients undergoing lumbar disc surgery. Over the past few years, additional evidence has accumulated to support the hypothesis that pre-emptive analgesia using a variety of agents (e.g., opiates, local anesthetics, NSAIDs) prolongs the time to first request for analgesics, reduces post-operative pain intensity, or decreases post-operative analgesic requirements among patients undergoing inguinal herniorraphy (Tverskoy et al. 1990), oral surgery (Tuffin et al. 1989; Hutchison et al. 1990; Campbell et al. 1990), tonsillectomy (Jebeles et al. 1991), abdominal surgery (Mogensen et al. 1992), orthopedic surgery (Ringrose and Cross 1984; McGlew et al. 1991), lower limb amputation (Mann and Bisset 1983; Bach et al. 1988) and thoracotomy (Katz et al. 1992).

Tverskoy et al. (1990) clearly demonstrated the benefits of pre-incisional blockade on post-operative pain. Patients who were undergoing inguinal herniorraphy received either general anesthesia alone, general anesthesia plus subcutaneous and intramuscular injections of bupivacaine prior to surgical incision, or spinal bupivacaine administered pre-operatively. All patients received the same regimen of post-operative analgesics. Twenty-four and 48 hours after surgery, post-operative incisional pain, movement-associated pain, and pain induced by pressure applied to the surgical wound were all significantly lower in the two groups that had received bupivacaine prior to surgical incision compared to patients that received general anesthesia alone.

Recently, a number of well-controlled, double-blind studies have also shown that pre-operative administration of NSAIDs by a variety of routes reduces post-operative pain long after the clinical duration of action of the NSAIDs. Campbell et al. (1990) found that intravenous diclofenac administered before tooth extraction resulted in less post-operative pain the day after surgery when compared with pretreatment using intravenous fentanyl or a placebo. Similarly, Hutchison et al. (1990) reported that, compared to patients pre-treated with a placebo, significantly fewer patients who received orally administered piroxicam before tooth extraction required supplemented post-operative analgesics, and their time to first post-operative analgesic request was longer. McGlew et al. (1991) demonstrated that on days 1 to 3 after spinal surgery, post-operative pain scores and opiate consumption were significantly lower among patients who had received indomethacin suppositories compared with placebo suppositories one hour before surgery.

Taken together, these studies demonstrate that opiate premedication, regional local anesthesia, spinal anesthesia, or systemic NSAIDs administered before incision are more effective than placebo or no treatment controls. The implication of these studies for clinical pathological pain is that changes in central neural function that are induced by surgery alter subsequent perception in such a way that nociceptive inputs from the surgical wound may be perceived as more painful (hyperalgesia) than they would otherwise have been, and innocuous inputs may give rise to frank pain (allodynia).

However, these early studies on the prevention of post-operative pain with pre-operative analgesics did not compare the pretreatment with the effects of the same treatments administered after surgery (McQuay 1992). Demonstrating that pretreatment with analgesics, but not a placebo, lessens pain and decreases post-operative analgesic requirements at a time when the agents are no longer clinically active indicates that the central component of post-operative pain can be prevented or pre-empted. In the absence of a post-incisional or post-operative treatment condition, it is not possible to determine the separate contributions of factors associated with the intra-operative versus the post-operative period to the enhanced post-operative pain experience. It may be that analgesic pretreatments reduce the development of local inflammation, a potential peripheral factor that could contribute to post-operative pain, rather than inhibiting central sensitization induced by noxious inputs during surgery. This may be particularly important in the case of NSAIDs (Campbell et al. 1990; Hutchison et al. 1990; McGlew et al. 1991), which act primarily to reduce peripheral inflammation, but may also be important in the case of infiltration with local anesthetics (Tverskoy et al. 1990), since local anesthesia would also reduce peripheral inflammation that is dependent on the efferent functions of peripheral nerves (i.e., neurogenic inflammation). Altering the timing of administration of analgesic agents (i.e., before or after incision vs. before or after surgery) may provide clues to the specific intra-operative (e.g., incision, wound retraction) or post-operative (e.g., inflammation) factors that contribute to the central neural changes underlying the enhanced pain.

Recently, studies have been directed at identifying specific intra- and post-operative factors that may contribute to surgically induced post-operative pain and hyperalgesia by comparing the effects on post-operative pain of opiates or local anesthetic agents administered either before or after surgery (Rice et al. 1990; Dahl et al. 1992; Dierking et al. 1992; Ejlersen et al. 1992; Katz et al. 1992). Rice et al. (1990) found that the timing of a caudal block with bupivacaine relative to the start of surgery had no effect on post-operative pain in a pediatric population undergoing brief (30 min) ambulatory surgical procedures. Dierking et al. (1992) evaluated the effects of a local-anesthetic inguinal field block administered before or after inguinal herniorraphy on post-operative pain and analgesic consumption. They also found that the timing of the block relative to surgical trauma did not produce differences in post-operative pain or analgesic use. Similarly, Dahl et al. (1992) reported that post-operative pain and analgesic consumption did not depend on whether a 72-hour continuous infusion of epidural bupivacaine and morphine was started before incision or immediately after surgery, approximately 2.5 hours later.

In contrast, Ejlersen et al. (1992) reported that even though pre-incisional blockade was not associated with significantly less post-operative pain, fewer patients in the pre-incisional group, as opposed to a post-incisional group, required supplemental post-operative analgesics, and their demand for analgesics was delayed. In addition, Katz et al. (1992) demonstrated that pre-incisional treatment with epidural fentanyl in patients undergoing thoracotomy resulted in significantly lower pain scores six hrs after treatment when compared with a post-incisional treatment. The significant difference in pain intensity could not be explained by lingering plasma concentrations of fentanyl, which at the time of pain assessment were equally sub-therapeutic in both groups, or by PCA morphine consumption, which until this point was virtually identical in both groups. Also, between 12 and 24 hours after surgery, the control group self-administered more than twice the amount of morphine than the experimental group, a finding that parallels the studies by Woolf and Wall (1986a, b). Recent studies by Katz and his colleagues (1994, 1996) continue to find small but consistent effects of pre-emptive analgesia on several types of post-surgical pain.

Experimental Evidence of CNS Plasticity

Damage of peripheral tissue and injury to nerves typically produce persistent pain and hyperalgesia. Recent evidence indicates that hyperalgesia depends, in part, on central sensitization. Hyperalgesia to punctate mechanical stimuli, which develops after intradermal injection of capsaicin, is maintained even after anesthetizing the region where capsaicin was injected (LaMotte et al. 1991). However, if the skin region is anesthetized prior to capsaicin injection, cutaneous hyperalgesia does not develop. Furthermore, hyperalgesic responses to capsaicin can be prevented if the area of skin where the injection is made is rendered anesthetic by a proximal anesthetic block of the peripheral nerve that innervates it. Thus, for hyperalgesia to develop it is critical that initial inputs from the injury reach the CNS. However, once hyperalgesia is established, it does not need to be maintained by inputs from the injured peripheral tissue.

Further evidence for a central mechanism of hyperalgesia is suggested by clinical and experimental cases of referred pain and hyperalgesia. Referred pain appears to depend on neural mechanisms, since local anesthesia of the injured region blocks its expression (Robertson et al. 1947). Furthermore, the role of central neural mechanisms is supported by the observation that phrenic nerve stimulation causes referred shoulder pain even after the sectioning of all cutaneous nerves from the painful region of the shoulder (Doran and Ratcliffe 1954), and by the finding that the injection of hypertonic saline into intraspinous ligaments resulted in pain referred to a phantom arm (Harman 1948). It is possible that referred pain depends on the misinterpretation of inputs from an injured region whose axons also branch to the uninjured referred area, or alternatively that axons from the injured and referred regions converge on the same cells in the sensory pathway. If referred pain could be explained exclusively by convergence, then such pains would not provide clear evidence of central sensitization. However, evidence that referred pain is also in part dependent on CNS changes is provided by findings that referred pain and hyperalgesia spread to areas that do not share the same dermatome (Livingston 1943). For example, it has been shown that pain of cardiac origin is referred to sites as distant as the patient's ear (Brylin and Hindfelt 1984). The fact that pain and hyperalgesia can spread to areas far removed from the injured region implies that central changes, as opposed to convergence, are involved in the spread of hyperalgesia.

Furthermore, referred pain has often been found to spread specifically to sites of a previous injury. Henry and Montuschi (1978) describe a case where the pain of an angina attack was referred to the site of an old vertebral fracture. Similarly, Hutchins and Reynolds (1947) discovered that alterations in barometric pressure during high-altitude flights caused many of their patients to complain of pain localized to teeth that had been the site of previous painful stimulation (e. g., fillings, caries and extractions), in many cases years earlier. Reynolds and Hutchins (1948) were able to replicate this finding under controlled conditions. One week after damaged teeth were filled or extracted, pinprick of the nasal mucosa produced pain referred to the previously treated teeth. This phenomenon occurred among patients who had been treated under general anesthesia, but not under the influence of a local anesthetic block. Futhermore, in patients who had received bilateral dental treatment without a local anesthetic, subsequent blocks applied to one side permanently abolished the referred pain ipsilateral but not contralateral to the anesthetized side.

Behavioral and physiological studies in animals also demonstrate hyperalgesia or sensitization in response to stimulation of body regions that are at a distance from a cutaneous or deep tissue injury. Cutaneous (Woolf 1983) and deep (Woolf and McMahon 1985) tissue injury, as well as noxious electrical stimulation of cutaneous and muscle afferent nerves (Wall and Woolf 1984), produce an increase in the excitability of the ipsilateral and contralateral flexor efferent nerves in response to noxious mechanical stimulation of the hindpaw. Since the increased excitability in the contralateral flexor efferent nerve is maintained even after inputs from the injured paw are blocked by local anesthesia, the results suggest that central, not peripheral, changes underlie this effect. In this way, cutaneous hyperalgesia after injury may depend on central hypersensitivity that is produced by inputs from a peripheral injury, but does not need to be maintained by them. Behavioral studies indicate that the spread of hyperalgesia to the hindpaw contralateral to the paw that received a thermal injury is unaffected by either deafferentation or anesthetic blocks of the injured hindpaw following the injury, but is prevented if deafferentation or anesthetic block precedes the injury (Coderre and Melzack 1987). These data provide further evidence that peripheral injury can produce central changes, that are maintained even after the inputs from the injury are removed.

Prolonged sensory disturbances associated with tissue injury (secondary hyperalgesia and referred pain, as well as allodynia and persistent spontaneous pain) are believed to result from either a reduction in the threshold of nociceptors or an increase in the excitability of CNS neurons involved in pain transmission. Since there is a large body of evidence documenting the sensitization of peripheral receptors following noxious stimulation, a peripheral mechanism is usually held to be responsible for the hyperalgesia that develops after injury. However, recent experimental studies suggest that sensitization within the CNS also contributes significantly to this phenomenon. These studies indicate that following injury, noxious stimulation, or C-fiber afferent electrical stimulation, there is a sensitization of neurons in the dorsal horn of the spinal cord and other areas in the somatosensory pathway. This sensitization is reflected by increased spontaneous activity, reduced thresholds or increased responsivity to afferent inputs, and prolonged afterdischarges to repeated stimulation.

In addition to the sensitization and prolonged excitation of dorsal horn cells, noxious stimulation associated with tissue injury also produces an expansion of the receptive fields of dorsal horn neurons. Neurons in the dorsal horn of the spinal cord with receptive fields adjacent to a cutaneous heat injury expand their receptive fields to incorporate the site of injury (McMahon and Wall 1984). Similar receptive field expansions have been observed in spinal cord following mechanical, chemical, inflammatory and nerve injuries, as well as following the induction of polyarthritis and in response to electrical nerve stimulation (see Coderre et al. 1993). Receptive field expansions have also been observed in brainstem and thalamic neurons.

Implications for Treatment of Acute and Chronic Pain

Recent advances in our understanding of the mechanisms that underlie patholog-ical pain have important implications for the treatment of both acute and chronic pain. Since it has been established that intense noxious stimulation produces a sensitization of CNS neurons, it is possible to direct treatments not only at the site of peripheral tissue damage but also at the site of central changes. Further-more, it may be possible in some instances to prevent the development of central changes that contribute to pathological pain states. The fact that amputees are more likely to develop phantom limb pain if there is pain in the limb prior to amputation (Katz and Melzack 1990), combined with the finding that the inci-dence of phantom limb pains is reduced if patients are rendered pain-free by epi-dural blockade with bupivacaine and morphine prior to amputation (Bach et al. 1988), suggests that the development of neuropathic pain can be prevented by reducing the potential for central sensitization at the time of amputation. The evidence that post-operative pain is also reduced by premedication with regional and/or spinal anesthetic blocks and/or opiates (McQuay et al. 1988; Tverskoy et al. 1990; Katz et al. 1992) suggests that acute post-operative pain can also benefit from the blocking of the afferent barrage arriving within the CNS and the central sensitization it may induce. Whether chronic post-operative problems such as painful scars, post-thoracotomy chest-wall pain, and phantom limb and stump pain can be reduced by blocking nociceptive inputs during surgery remains to be determined. Furthermore, additional research is required to determine whether multiple-treatment approaches (involving local and epidural anesthesia, as well as pretreatment with opiates and anti-inflammatory drugs) that produce an effective blockade of afferent input may also prevent or relieve other forms of severe chronic pain, such as post-herpetic neuralgia and reflex sympathetic dys-trophy. It is hoped that a combination of new pharmacological developments, careful clinical trials, and an increased understanding of the contribution and mechanisms of noxious stimulus-induced neuroplasticity will lead to improved clinical treatment and prevention of pathological pain.

Acknowledgements

Supported by grant A7891 from the Natural Sciences and Engineering Research Council of Canada (NSERC) to R.M., grant MT-11045 from the Medical Research Council (MRC) of Canada and grant 900051 from Fonds de la Recherche en Santé du Québec to T.J.C., and Fellowships from MRC to J.K., and NSERC to A.L.V.

References

Albe-Fessard D, Lombard MD (1983) Use of an animal model to evaluate the origin of and protection against deafferentation pain. In: Bonica JJ, Lindblom U, Iggo A (eds) Advances in pain research and therapy, Vol 5. Raven Press, New York, pp 691–700

Asada H, Yasumo W, Yamaguchi Y (1990) Properties of hyperactive cells in rat spinal cord after peripheral nerve section. Pain (Suppl. 5):S22

Bach S, Noreng MF, Tjéllden NU (1988) Phantom limb pain in amputees during the first 12 months following limb amputation, after preoperative lumbar epidural blockade. Pain 33:297–301

Basbaum AI, Wall PD (1976) Chronic changes in the response of cells in adult cat dorsal horn following partial de-afferentation: the appearance of responding cells in a previously non-responsive region. Brain Res 116:181–204

Berger M, Gerstenbrand F (1981) Phantom illusions in spinal cord lesions. In: Siegfried J, Zimmermann M (eds) Phantom and stump pain. Springer, Berlin, pp. 66–73

Blumenkopf B, Lipman JJ (1991) Studies in autotomy: its pathophysiology and usefulness as a model of chronic pain. Pain 45:203–210

Brena SF, Sammons EE (1979) Phantom urinary bladder pain – Case report. Pain 7:197–201

Brylin M, Hindfelt B (1984) Ear pain due to myocardial ischemia, Am Heart J 107:186–187

Campbell WI, Kendrick R, Patterson C (1990) Intravenous diclofenac sodium. Does its administration before operation supress postoperative pain? Anaesthesia 45:763–766

Coderre TJ, Melzack R (1987) Cutaneous hyperalgesia: contributions of the peripheral and central nervous system to the increase in pain sensitivity after injury. Brain Res 404:95–106

Coderre TJ, Fisher K, Fundytus ME (1997) The role of ionotropic and metabotropic glutamate receptors in persistent nociception. In: Jensen TS, Turner JA, Wiesenfeld-Hallin (eds) Proceedings of the 8th World Congress of Pain, IASP Press, Seattle, pp. 259–275

Coderre TJ, Grimes RW, Melzack R (1986) Autotomy after nerve sections in the rat is influenced by tonic descending inhibition from locus coeruleus. Neurosci Lett 67:82–85

Coderre TJ, Katz J, Vaccarino AL, Melzack R (1993) Contribution of central neuroplasticity to pathological pain: review of clinical and experimental evidence. Pain 52:259–285

Coderre TJ, Vaccarino AL, Melzack R (1990) Central nervous system plasticity in the tonic pain response to subcutaneous formalin injection. Brain Res 535:155–158

Crile GW (1913) The kinetic theory of shock and its prevention through anoci-association (shockless operation). Lancet (ii) 7–16

Dahl JB, Hansen BL, Hjortso NC, Erichsen CJ, Moiniche S, Kehlet H (1992) Influence of timing on the effect of continuous extradural analgesia with bupivacaine and morphine after major abdominal surgery. Br J Anaesth 69:4–8

Dennis SG, Melzack R (1979) Self-mutilation after dorsal rhizotomy in rats: Effects of prior pain and pattern of root lesions. Exp Neurol 65:412–421

Dickenson AH, Sullivan AF (1987a) Evidence for a role of the NMDA receptor in the frequency dependent potentiation of deep rat dorsal horn nociceptive neurones following C fibre stimulation. Neuropharmacology 26:1235–1238

Dickenson AH, Sullivan AF (1987b) Subcutaneous formalin-induced activity of dorsal horn neurons in the rat: differential response to an intrathecal opiate administration pre or post formalin. Pain 30:349–360

Dierking GW, Dahl JB, Kanstrup J, Dahl A, Kehlet H (1992) Effect of pre- vs postoperative inguinal field block on postoperative pain after herniorraphy. Br J Anaesth 68:344–348

Doran FSA, Ratcliffe AH (1954) The physiological mechanism of referred shoulder-tip pain. Brain 77:427–434

Dorpat TL (1971) Phantom sensations of internal organs. Comp Psychiat 12:27–35

Dougherty PM, Garrison CJ, Carlton SM (1992) Differential influence of local anesthesia upon two models of experimentally induced peripheral mononeuropathy in rat. Brain Res 570:109–115

Dubuisson D, Dennis SG (1977) The formalin test: a quantitative study of the analgesic effects of morphine, meperidine, and brain stimulation in rats and cats. Pain 4:161–174

Ejlersen E, Bryde-Anderson H, Eliasen K, Mogensen T (1992) A comparison between preincisional and postincisional lidocaine infiltration and postoperative pain. Anesth Analg 74:495–498

Gloyne HF (1954) Psychosomatic aspects of pain. Psychoanal. Rev. 41:135–159

Go VLW, Yaksh TL (1987) Release of substance P from the cat spinal cord. J Physiol 391:141–167

Hardy JD, Wolff HG, Goodell H (1950) Experimental evidence on the nature of cutaneous hyperalgesia. J Clin Invest 29:115–140

Harman JB (1948) The localization of deep pain. Br Med J 1:188

Henry JA, Montuschi E (1978) Cardiac pain referred to site of previously experienced somatic pain. Br Med J 9:1605–1606

Hodge CJ, Apkarian AV, Owen MP, Hanson BS (1983) Changes in the effects of stimulation of locus coeruleus and nucleus raphe magnus following dorsal rhizotomy. Brain Res 288:325–329

Howe JF, Loeser JD, Calvin WH (1977) Mechanosensitivity of dorsal root ganglia and chronically injured axons: a physiological basis for the radicular pain of nerve root compression. Pain, 3:25–41

Hutchins HC, Reynolds OE (1947) Experimental investigation of the referred pain of aerodontalgia. J Dent Res 26:3–8

Hutchison SL, Crofts SL, Gray IG (1990) Preoperative piroxicam for postoperative analgesia in dental surgery. Br J Anaesth 65:500–503

Jebeles JA, Reilly JS, Guitierrez JF, Bradley EL, Kissin I (1991) The effect of pre-incisional infiltration of tonsils with bupivacaine on the pain following tonsillectomy under general anesthesia. Pain 47:305–308

Jensen TS, Rasmussen P (1989) Phantom pain and related phenomena after amputation. In: Wall PD, Melzack R (eds.) Textbook of pain, 2nd ed. Livingstone Churchill, Edinburgh, pp 508–521

Jensen TS, Krebs B, Nielsen J, Rasmussen P (1985) Immediate and long-term phantom pain in amputees: incidence, clinical characteristics and relationship to pre-amputation pain. Pain 21:267–278

Katz J, Melzack P (1990) Pain 'memories' in phantom limbs: review and clinical observations. Pain 43:319–336

Katz J, Vaccarino AL, Coderre TJ, Melzack R (1991) Injury prior to neurectomy alters the pattern of autonomy in rats: behavioral evidence of central neural plasticity. Anesthesiology 75:876–883

Katz J, Kavanagh BP, Sandler AN, Nierenberg H, Boylan JF, Shaw BF (1992) Pre-emptive analgesia: clinical evidence of neuroplasticity contributing to postoperative pain. Anesthesiology 77:439–446

Katz J, Claireux M, Kavanagh BP, Rogers S, Nierenberg H, Redahan C, Sandler AN (1994) Pre-emptive lumbar epidural anaesthesia reduces postoperative pain and patient-controlled morphine consumption after lower abdominal surgery. Pain 59:395–403

Katz J, Jackson M, Kavanaugh BP, Sandler AN (1996) Acute pain after thoracic surgery predicts long-term post-thoractomy pain. Clin J Pain 12:50–55

Kenshalo DR Jr, Leonard RB, Chung JM, Willis WD (1982) Facilitation of the responses of primate spinothalamic cells to cold and mechanical stimuli by noxious heating of the skin. Pain 12:141–152

Kiss IE, Kilian M (1992) Does opiate premedication influence postoperative analgesia? A prospective study. Pain 48:157–158

Kroner K, Krebs B, Skov J, Jorgensen HS (1989) Immediate and long-term breast syndrome after mastectomy: incidence, clinical characteristics and relationship to pre-mastectomy breast pain. Pain 36:327–337

LaMotte RH, Shain CN, Simone DA, Tsai E-FP (1991) Neurogenic hyperalgesia: psychophysical studies of underlying mechanisms. J Neurophysiol 66:190–211

Lenz FA, Kwan HC, Dostrovsky JO, Tasker RR (1989) Characteristics of the bursting pattern of action potential that occurs in the thalamus of patients with central pain. Brain Res 496:357–360

Lenz FA, Tasker RR, Dostrovsky JO, Kwan HC, Gorecki J, Hirayama T, Murphy JT (1987) Abnormal single-unit activity recorded in the somatosensory thalamus of a quadriplegic patient with central pain. Pain 31:225–236

Leriche R (1947) A propos des algies des amputées. Mém Acad Chir 73:280–284

Livingston WK (1943) Pain mechanisms. MacMillan, New York

MacKenzie J (1893) Some points bearing on the association of sensory disorders and visceral diseases. Brain 16:321–354

Macon JB (1979) Deafferentation hyperactivity in the monkey spinal trigeminal nucleus: neuronal responses to amino acid iontophoresis. Brain Res 161:549–554

Mann RAM, Bisset WIK (1983) Anaesthesia for lower limb amputation. A comparison of spinal analgesia and general anaesthesia in the elderly. Anaesthesiology 38:1185–1191

Markus H, Pomeranz B, Krushelnycky D (1984) Spread of saphaneous somatotopic projection map in spinal cord and hypersensitivity of the foot after chronic sciatic denervation in adult rat. Brain Res 296:27–39

McGlew IC, Angliss DB, Gee GJ, Rutherford A, Wood ATA (1991) A comparison of rectal indomethacin with placebo for pain relief following spinal surgery. Anaesth Intensive Care 19:40–45

McMahon SB, Wall PD (1984) Receptive fields of rat lamina 1 projection cells move to incorporate a nearby region of injury. Pain 19:235–247

McQuay HJ (1992) Pre-emptive analgesia. Brit J Anaesth 69:1–3

McQuay HJ, Carroll D, Moore RA (1988) Post-operative orthopaedic pain – the effect of opiate premedication and local anaesthetic blocks. Pain 33:291–295

Melzack R, Wall PD (1965) Pain mechanisms: a new theory. Science 150:971–979

Melzack R, Wall PD (1996) The challenge of pain. Penguin, London (Updated Second Edition)

Minski L (1943) Psychological reactions to injury. In: Doherty WB, Runes BB (eds) Rehabilitation of the war injured. Philosophical Library, New York, pp 115–122

Mogensen T, Bartholdy J, Sperling K, Ibsen M, Eliasen K (1992) Preoperative infiltration of the incisional area enhances postoperative analgesia to a combined low-dose epidural bupivacaine and morphine regime after upper abdominal surgery. Reg Anesth 17 Suppl. 74

Nathan PW (1962) Pain traces left in the central nervous system. In: Keele CA, Smith R (eds) The assessment of pain in man and animals. Livingstone, Edinburgh, pp 129–134

Nathan PW (1985) Pain and nociception in the clinical contex. Phil Trans Royal Soc London 308:219–226

Noordenbos W, Wall P (1981) Implications of the failure of nerve resection and graft to cure chronic pain produced by nerve lesions. J Neurol Neurosurg Psychiat 44:1068–1073

Oliveras JL, Guilbaud G, Besson JM (1979) A map of serotonergic structures involved in stimulation produced analgesia in unrestrained freely moving cats. Brain Res 164:317–322

Ovesen P, Kroner K, Ornsholt J, Bach K (1991) Phantom-related phenomena after rectal amputation: prevalence and clinical characteristics. Pain 44:289–291

Pare D, Steriade M, Deschenes M, Bouhassira D (1990) Prolonged enhancement of anterior thalamic synaptic responsiveness by stimulation of a brain-stem cholinergic group. J Neurosci 10:20–33

Reynolds OE, Hutchins HC (1948) Reduction of central hyper-irritability following block anesthesia of peripheral nerve. Am J Physiol 152:658–662

Rice LJ, Pudimat MA, Hannallah RS (1990) Timing of caudal block placement in relation to surgery does not affect duration of postoperative analgesia in paediatric ambulatory patients. Can J Anaesth 37:429–431

Rinaldi PC, Young RF, Albe-Fessard D, Chodakiewitz J (1991) Spontaneous neuronal hyperactivity in the medial and intralaminar thalamic nuclei of patients with deafferentation pain. J Neurosurg 74:415–421

Ringrose NH, Cross MJ (1984) Femoral nerve block in knee joint surgery. Am J Sports Med 12:398–402

Robertson S, Goodell H, Wolff HG (1974) Headache: the teeth as a source of headache and other head pain. Arch Neurol Psychia 57:277–291

Segal M, Sandberg D (1977) Analgesia produced by electrical stimulation of catecholamine nuclei in the rat brain. Brain Res 123:369–372

Seltzer Z, Beilin BZ, Ginzburg, R, Paran Y, Shimko T (1991) The role of injury discharge in the induction of neuropathic pain behavior in rats. Pain 46:327–336

Sinclair JG, Lo GF (1986) Morphine, but not atropine, blocks nociceptor-driven activity in rat dorsal hippocampal neurones. Neurosci Lett 68:47–50

Szasz TS (1949) Psychiatric aspects of vagotomy: IV. Phantom ulcer pain. Arch Neurol Psychiat 62:728–733

Tasker RR (1989) Stereotactic surgery. In: Wall PD, Melzack R (eds) Textbook of pain, 2nd edition. Livingstone Churchill, Edinburgh, pp 840–855

Tuffin JR, Cunliffe DR, Shaw SR (1989) Do local analgesics injected at the time of third molar removal under general anesthesia reduce significantly postoperative analgesic requirements? A double-blind controlled trial. Br J Oral Maxillofacial Surg 27:27–32

Tverskoy M, Cozacov C, Ayache M, Bradley EL, Kissin I (1990) Postoperative pain after inguinal herniorraphy with different types of anesthesia. Anesth Anal 70:29–35

Vaccarino AL, Melzack R (1992) Temporal processes of formalin pain: differential role of the cingulum bundle, fornix pathway and medial bulboreticular formation. Pain 44:257–271

Vinogradova OS (1975) Functional organization of the limbic system in the process of registration of information: facts and hypothesis. In: Isaacson RL, Pribram KH (eds) The hippocampus, Vol 2. Plenum Press, New York, pp 3–69

Wall PD, Devor M (1981) The effect of peripheral nerve injury on dorsal root potentials and on transmission of afferent signals into the spinal cord. Brain Res 209:95–111

Wall PD, Devor M (1983) Sensory afferent impulses originate from the dorsal root ganglia as well as from the periphery in normal and nerve injured rats. Pain 17:321–339

Wall PD, Gutnik M (1974) Properties of afferent nerve impulses originating from a neuroma. Nature 248:740–743

Wall PD, Woolf CJ (1984) Muscle but not cutaneous C-afferent input produces prolonged increases in the exitability of the flexion reflex in the rat. J Physiol (Lond) 356:443–458

Wall PD, Scadding JW, Tomkiewicz MM (1979) The production and prevention of experimental anesthesia dolorosa. Pain 6:179–182

Wallgren GR (1954) Phantom experience at spinal anaesthesia. Ann Chir et Gynaec Fenniae 43 (Suppl):486–500

Woolf CJ (1983) Evidence for a central component of post-injury pain hypersensitivity. Nature 306:686–688

Woolf CJ, McMahon SB (1985) Injury-induced plasticity of the flexor reflex in chronic decerebrate rats. Neurosci 16:395–404

Woolf CJ, Wall PD (1982) Chronic peripheral nerve section diminishes the primary afferent A-fibre mediated inhibition of rat dorsal horn neurons. Brain Res 242:77–85

Woolf CJ, Wall PD (1986a) Morphine-sensitive and morphine-insensitive actions of C-fibre input on the rat spinal cord. Neurosci Lett 64:221–225

Woolf CJ, Wall PD (1986b) Relative effectiveness of C primary afferent fibers of different origins in evoking a prolonged facilitation of the flexion reflex in the rat. J Neurosci 6:1433–1442

Auditory Cortical Plasticity and Sensory Substitution

J. P. Rauschecker[*]

Summary

Early visual deprivation in cats and ferrets leads to crossmodal compensation on both the behavioral and neurobiological levels (Rauschecker 1995). Visually deprived animals can localize sounds in space with greater precision. Correspondingly, neurons in areas of cortex normally activated by visual stimuli are now activated by sound. Not only are the cortical areas in the parietal lobe subserving auditory function expanded; neurons in these areas are also more sharply tuned to auditory spatial location.

In blind humans, strikingly similar results are seen with functional neuroimaging techniques. Positron emission tomography (PET) shows that activation zones in the inferior parietal lobules that are associated with auditory spatial processing are vastly expanded into occipital areas. The same change is found to a much lesser degree for subjects who acquired their blindness later in life. Both sets of data demonstrate the vast capacity of the cerebral cortex to reorganize itself depending on the demands of altered environmental conditions.

Introduction

Our knowledge about cortical plasticity stems mainly from studies of the visual and somatosensory systems. Oddly enough, the auditory system has long been considered by many to be more hard-wired. This view undoubtedly originated from the concentration of work on the peripheral auditory system, which, like other sensory peripheries, may indeed be more hard-wired than its central representations. Recently, studies of the central auditory system have gained more interest and, consequently, interest in auditory plasticity is also on the rise. My own contribution to this field so far comes mainly from studies of auditory cortical reorganization after early blindness, i.e., a form of crossmodal plasticity in which reorganization and expansion of auditory areas in the cerebral cortex are fostered by the loss of input in another sensory modality, vision. I will include, however, other examples of auditory plasticity in my discussion of these findings.

[*] Georgetown Institute for Cognitive and Computational Sciences, Georgetown University Medical Center, Washington, DC 20007, USA

J. Grafman / Y. Christen (Eds.)
Neuronal Plasticity:
Building a Bridge from the Laboratory to the Clinic
© Springer-Verlag Berlin Heidelberg New York 1999

Reorganization of Auditory Representations After Visual Deprivation in Animal Models

Anecdotal evidence suggests that blind musicians may acquire their superior capacity for music (or audition in general) from an early vision loss. While such widely held beliefs have been around for some time, a careful review of the literature suggests that the results are not so clear. About the same number of studies are found pointing to a compensation for the early vision loss by increased capacities in audition and touch as studies that claim just the opposite, namely, that an early vision loss can lead to a general impairment of sensory function because vision guides the development of other sensory systems. After such a review of the clinical literature, astonishingly, one ends up with two almost diametrically opposed hypotheses (Rauschecker 1995). The "compensatory plasticity hypothesis" states that sensory deprivation in one modality leads to compensation or substitution of the lost sense by non-deprived modalities. Specifically it has been postulated that, for blind people capacities are developed in their remaining senses that may exceed those of sighted individuals. By contrast, what one could call a "general degradation hypothesis" states that sensory deprivation leads to general impairment of brain development of cognitive function. Specifically, blindness leads to degradation of auditory localization capacities because vision is needed to calibrate space and visualization is needed for auditory and tactile form perception.

I had always found the degradation hypothesis fairly hard to swallow due to observations I had made over the years in our colony of visually deprived cats who were lid-sutured shortly after birth and thus never had any pattern vision. As noted by Wiesel and Hubel (1965) in some of their early papers, such animals display capacities in their behavior that are almost impossible to distinguish from normal animals. So, early on in these studies we decided to measure quantitatively the auditory localization behavior of visually deprived cats. The study, which was conducted by my graduate student Ulla Kniepert, was done in the following way (Rauschecker and Kniepert 1994). An arena was placed in a sound-attenuated chamber that contained eight speakers at equal distances in azimuth (Fig. 1 A). The cat was placed in the start box in the middle of the arena and was trained to go toward the randomly activated speakers in order to get a food reward. The cat had to identify the speaker location as precisely as possible, and its sound localization performance was measured. Figure 1 B shows how the sound localization error, measured as the mean variance, depends on the speaker position. Speaker position 1, which is straight ahead of the animal, is associated with the smallest localization error. Speaker position 5, which is straight behind the animal, usually leads to the largest localization error, and speaker positions in between gradually generate intermediate values of localization error. This finding corresponds exactly to human psychophysics, where sound localization also depends on position in space (Blauert 1996). The main point of Figure 1 B, however, is the difference between the normally sighted and the visually deprived animals. Every speaker position's sound localization error was smaller in blind

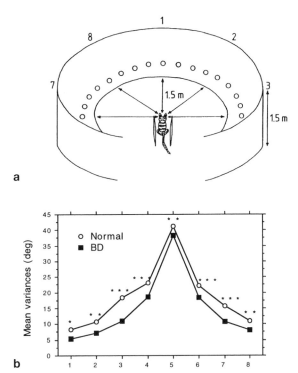

Fig. 1. Sound localization performance in normal and blind cats (after Rauschecker and Kniepert 1994; Rauschecker 1995).
A, testing apparatus with speakers in free field.
B, sound localization error is significantly smaller in blind cats (BD, binocularly deprived) at all speaker positions, but most significantly so for lateral and rear positions.

animals than in normal animals and the difference, on average, was most significant for lateral and rear positions. Again, this result corresponds to that in blind humans (Rice et al. 1965).

Having established that blind animals are indeed at least as good and almost certainly superior in sound localization ability as sighted animals, we tried to find the neural substrate for these improvements by performing electrophysiological recordings from the brains of these animals. We had received hints from neuroanatomical studies conducted in collaboration with Andreas Aschoff (Rauschecker and Aschoff 1987) that an area in the cat's parietal cortex which projects heavily to the superior colliculus of the midbrain might be a good candidate structure for changes in the brain that subserved the improved auditory localization capacity. This region of the brain is called the anterior ectosylvian sulcus (AES; Fig. 2A) and, in close vicinity, receives input from three sensory modalities: somatosensory, auditory, and visual. The visual region (AEV or EVA), which was first described by Mucke et al. (1982) and by Olson and Graybiel (1983), contains visual receptive fields that respond to visual stimuli of a moving type. The neighboring auditory regions contain neurons tuned to spatially discrete sounds (Rauschecker et al. 1993; Korte and Rauschecker 1993). The anterior ectosylvian region was explored by single unit recording in both normal and blind animals. The examples in Figure 2B (top) show penetrations in normal animals; they always start out dorsally in a portion of auditory cortex, the anterior

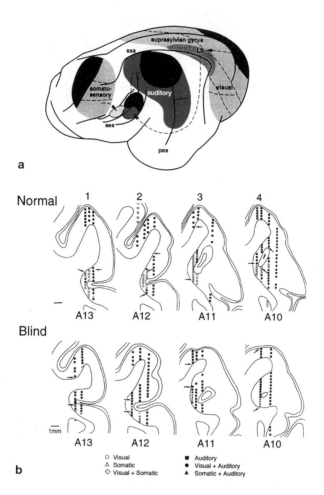

Fig. 2. Electrophysiological single unit recordings from the anterior ectosylvian cortex cats (after Korte and Rauschecker 1993; Rauschecker 1995). **A**, lateral view of cat brain with anterior ectosylvian sulcus (aes). **B**, track reconstructions through the anterior ectosylvian sulcus of normal and blind cats. Normally purely visual neurons in the fundus of aes are replaced by auditory-responsive cells. pes, posterior ectosylvian sulcus; sss, suprasylvian sulcus.

auditory field (AAF), and then end up in a region (AEV) in the fundus of the anterior ectosylvian sulcus that is purely visually driven. By contrast, penetrations in blind animals (Fig. 2B, bottom) lack this visual representation in the fundus of AES. Instead this portion of cortex contains neurons that are auditory or somatosensory. At best, these neurons are driven by both visual and auditory stimuli, but none of the cells that were found there in any of the blind animals were purely visual. I should emphasize that semi-chronic recording was used in all of these animals, so the complete AES region was explored in every cat. This excludes the possibility that we had introduced a sampling bias by incomplete analysis of the region. All neurons were characterized with stimuli in all three sensory modalities. The remarkable thing is that in the fundus of the AES region of blind animals we do not see large numbers of unresponsive neurons due to the loss of visual input. Instead, we find neurons that had changed their response properties from visual to auditory or somatosensory. Figure 3 shows a summary of these results. With visual deprivation, the number of auditory-responsive neurons goes up significantly while at the same time the number of purely visual neurons goes down (Fig. 3 A). The number of bimodal neurons also increases significantly, whereas the proportion of nonresponsive neurons remains unchanged. Bimodal neurons increase in all cetegories but mainly in those that have an auditory input to them (Fig. 3 B).

An increase in the number of auditory-responsive neurons in the AES region is interesting but may not by itself provide evidence for the neurobiological basis of the improved sound localization ability that we found in blind animals. Therefore, we attempted to analyze a more quantitative, fine-level parameter, namely spatial tuning of single neurons in AES. Examples given in Figure 4 demonstrate that, in normal animals, spatially tuned neurons are indeed found that respond

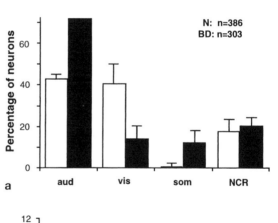

Fig. 3. Distribution of neurons with different sensory modalities in AES of normal (N) and blind (BD) cats (from Rauschecker and Korte 1993). **A**, significantly more auditory responsive neurons are found in the AES region with no increase in the number of unresponsive cells (NCR). **B**, the number of bimodal neurons is increased, particularly those with an auditory input.

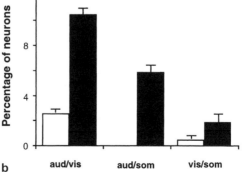

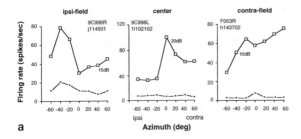

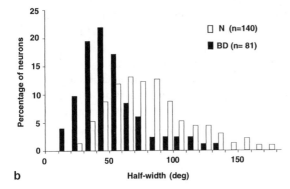

Fig. 4. Auditory spatial tuning of single neurons in AES as a function of visual experience. **A,** examples of spatial tuning curves. **B,** distribution of auditory spatial tuning in normal (N) and binocularly deprived (BD) animals. Spatial tuning is significantly sharper in blind animals.

best to a particular location and less so to others. In blind animals, the sharpness of tuning for the sound location in azimuth is greatly increased. The experiment was conducted in the following way. Speakers were put up in free field, and sounds were presented randomly at different locations. The firing rate of neurons in the gas-anesthetized animals was measured as a function of the location of the speaker. Peak firing rates were then assembled into a spatial tuning curve. Different intensities were also tried in the same neurons, and the peak of this best azimuth was always found to be the same regardless of intensity.

A comparison of the spatial tuning sharpness was performed either on the basis of a spatial tuning index or of the half-width at half height, which was calculated for neurons both in the normal and the blind group. The spatial tuning index is simply a ratio of the best and the least response in different locations: if a neuron responds equally to all locations, the spatial tuning index is close to 1. If the response is much better in one location than in another, it would be closer to 0. Neurons with spatial tuning indices below 0.5 were considered "directional." The number of directional units is much larger in the blind animals, approaching 90%, whereas in normal animals only about 50% reached this criterion. Half-width was determined as the tuning width (in degrees) at half maximal height of the spatial tuning curve (Fig. 4B).

In summary, we are dealing with an expansion of non-visual representations in parietal cortex of blind cats as a result of visual deprivation. A purely visual representation in the anterior ectosylvian cortex not only shrinks but almost dis-

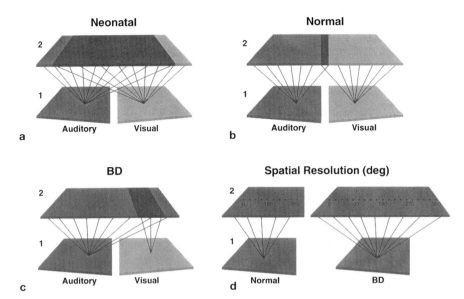

Fig. 5. Schematic representation of auditory substitution (Rauschecker 1995). **A,** initial condition in newborn cats with large region of multimodal overlap. **B,** with normal experience, the overlap region is reduced by activity-dependent competition. **C,** visual deprivation (BD) provides a competitive advantage to auditory input. **D,** sharpened spatial tuning of auditory units can be explained within the same model (see text).

appears, and neighboring regions that are auditorily driven or driven by somato-sensory input expand at the expense of the visual representation. In a simple computational model, one could imagine starting out from a state in neonatal animals where there is great overlap between auditory and visual inputs in a multimodal convergence area (Fig. 5A). During normal development, this overlap region largely disappears, and areas dominated by auditory or visual inputs emerge (Fig. 5B). In blind animals, due to the lack of visual input, the auditory region has a better chance of expanding and establishing itself in the multimodal target region, and therefore one would find the results that we have shown experimentally (Fig. 5C). The same model can explain the improvement in spatial tuning of neurons in auditory cortex because an increased area, with a larger number of auditory neurons, leads to a smaller angle of representation by each single neuron within this map (Fig. 5D).

In our lab, Peter Henning has found that neurons in primary auditory cortex (A1) may also show some improvement of auditory spatial tuning, whereas the elevation tuning of such neurons has not been found to improve (Henning and Rauschecker 1995). However, we have found another mechanism that may also improve the ability of blind cats to localize sound in elevation. When startled by a sudden sound that appears in the room, blind cats display a very characteristic form of behavior that normal cats never show. First they orient toward the sound source and then start scanning movements in elevation around the center of the presumed sound location, involving both head and pinnae in a coordinated fashion (Henning and Rauschecker 1993). We do not know at this time where the programming of these scanning movements occurs, but we assume that it is triggered by an interaction of cortical and midbrain mechanisms.

Reorganization of Auditory Representations in Blind Humans

In the second part of my chapter I would like to present results of sound localization studies in congenitally blind humans, which were performed at the National Institutes of Health in collaboration with Robert Weeks in the laboratory of Mark Hallett and with the help of several other co-workers (Aziz-Sultan et al. 1997; Weeks et al. 1997). Our laboratory provided the auditory stimuli, which were programmed on a computer system (Tucker Davis Technologies) and presented via headphones. The system produced a virtual auditory space environment by means of presenting appropriate combinations of interaural time and intensity differences as well as spectral cues based on head-related transfer functions. Subjects were tested in different conditions. In the first condition, they had to listen only to a tone that was presented straight ahead of them. In a second condition, they had to localize a sound that was presented at random in different azimuth positions, and they had to judge the position of a second sound, presented shortly after the first one, in relation to the position of the previous one. In the third condition, subjects had to make the same judgment but actually indicate the position of the sound with a joystick they had in their hand.

Contrast one, the comparison of passive listening condition one versus rest, in sighted subjects shows activation in primary auditory cortex and surrounding areas in the superior temporal region. When the activaton of condition two (delayed matching auditory localization task) is compared to rest we found that, in addition to prefrontal regions needed presumably to keep the position of the first sound in working memory, inferior-parietal regions were distinctly activated bilaterally, with a slight bias toward the right hemisphere. When the auditory localization condition in blind subjects was compared to rest, the same inferior parietal regions were active as in normal subjects; however, there was massive activation of occipital areas in addition to the parietal regions. This finding was the same whether the delayed matching auditory localization task was considered or the active motor task involving the joystick was used. Typical activation involved mainly Brodmann areas 18 and 19, and to a lesser extent (or not at all) the primary visual cortex, area 17. Between-group analysis showed that the differences in the occipital areas were highly significant, both for the comparison of auditory localization versus rest and for auditory localization versus passive listening.

In summary, not only do we find distinct activation of inferior parietal regions by virtual auditory space stimuli, which suggests an involvement of the inferior parietal region in auditory spatial processing, but also most significantly in the present context, we find an increase in activity starting out from this parietal region emanating into occipital areas 18 and 19, which are normally used for visual tasks.

Reorganization of Auditory Representations in Deaf Humans

In a study complementary to the previous one we studied cortical reorganization during early development in deaf subjects. This was a collaboration that involved Helen Neville, then at UCSD, and her co-workers Daphne Bavelier and David Corina, as well as Peter Jezzard and Bob Turner, then at NIH, and a number of other colleagues (Neville et al. 1998). The studies were performed using functional magnetic resonance imaging (fMRI) on a 4-Tesla Bruker magnet at the NIH. The task was the same for hearing and deaf subjects. While lying in the magnet, the subjects were presented visual stimuli consisting of written English or of movies of American Sign Language (ASL). The result was that, in hearing subjects reading written English, large portions of the superior and middle temporal regions, including Wernicke's area, were activated in addition to some visual and frontal regions. By contrast, in deaf subjects reading American Sign Language, activation of the left temporal regions was less pronounced, and homologous areas in the right hemisphere were activated in addition. Furthermore, in the deaf subjects, parietal areas in the right hemisphere were activated that do not light up in hearing subjects under the same conditions. This finding demonstrates the capacity of the cerebral cortex for crossmodal reorganization after early auditory deprivation.

In conclusion from both animal and human studies, we can confirm the existence of compensatory plasticity and sensory substitution in the cerebral cortex. The loss of one sensory modality leads to reorganization of cortical representations in the remaining modalities. Not only does an expansion occur of those representations that are more active during development, but, from the animal studies we can additionally conclude that single neuron response properties are sharpened by more intense sensory experience. This is reflected by behavioral improvements both in animals and in humans with this type of restricted sensory experience. Whether there needs to be actual competition between these sensory modalities for this improvement to occur is still unclear. In the case of a competitive mechanism, the nondeprived modalities would profit at the expense of representations that are not adequately stimulated. Alternatively, increased training in one modality providing increased activity would lead to the expansion even in the presence of the other modality.

This process could be further intensified by attentional mechanisms. However, even if attention helps to guide the initial improvement and expansion in other modalities, hard-wired changes will ultimately result, as we find in our animal studies. In these experiments, physiological parameters were measured under general anesthesia, which excludes simple modulatory factors. Therefore, by analogy, the apparent expansion of nondeprived modalities in the human imaging studies, which could have potentially been caused by temporary attentional modulation, can also most likely be explained by permanent changes in cortical circuitry.

Outlook on other Forms of Auditory Plasticity

The capacity of auditory cortex to reorganize itself under the above conditions gives hope that it will prove equally plastic in other situations. Perhaps the most important practical implication of auditory cortical plasticity is with regard to the effectiveness of cochlear implants. Much effort has been devoted to developing coding mechanisms that best adapt the implant as a transducer to the signals of the auditory nerve. Little thought has been given, on the other hand, to the plasticity of the ultimate receiver in the brain, the auditory cortex, which has to deal with processing the incoming nerve signals for auditory perception. The surprising success of the rather crude signals provided by some of the cochlear implants demonstrates clearly that the central auditory system, with its immense plasticity, can adapt to a new, if rudimentary code. Future efforts should be devoted to the fact that cortical structures are known to be most plastic early in life, so implants may have to be applied in neonates as soon as possible after birth.

A similar argument applies to considerations about the effects of chronic otitis media on the development of the central auditory system. The resulting auditory deprivation, if not prevented from recurring within the critical periods for auditory cortical development, may lead to permanent damage of higher auditory processing capacities, including speech perception. The most important period happens before the age of two, around 18–21 months, when children go through a "language spurt" resulting in a rapid expansion of their active vocabulary (Nobre and Plunkett 1997).

There are important implications for auditory plasticity in an even wider sense. The necessity for the auditory cortex to receive adequate stimulation during early development implies that the lack of stimulation can lead to severe impairments in later language learning and thus for learning in general. Such effects have indeed been demonstrated (Merzenich et al. 1996; Tallal et al. 1996; Kraus et al. 1996), and one of the most important goals for society remains to ascertain a fuller understanding not only of the causes of hearing impairment but also of their treatment by making full use of the malleability of the auditory system early and later in life.

References

Aziz-Sultan A, Weeks RA, Tian B, Cohen LG, Rauschecker JP, Hallett M (1997) Auditory localisation demonstrates cross modal plasticity in congenitally blind subjects. Neurology 48:S30.004

Blauert J (1996) Spatial Hearing, 2nd edn. Cambridge, MA, MIT Press

Henning P, Rauschecker JP (1993) Vertical scanning movements of head and pinnae in visually deprived cats. Soc Neurosci Abstr 19:164

Henning PT, Rauschecker JP (1995) Organization of spatial tuning columns in the auditory cortex of normal and visually deprived cats. Soc Neurosci Abstr 21:668

Korte M, Rauschecker JP (1993) Auditory spatial tuning of cortical neurons is sharpened in cats with early blindness. J Neurophysiol 70:1717–1721

Kraus N, McGee TJ, Carrell TD, Zecker SG, Nicol TG, Koch DB (1996) Auditory neurophysiologic responses and discrimination deficits in children with learning problems. Science 273:971–973

Merzenich MM, Jenkins WM, Johnston P, Schreiner C, Miller SL, Tallal P (1996) Temporal processing deficits of language-learning impaired children ameliorated by training. Science 271:77–81

Mucke L, Norita M, Benedek G, Creutzfeldt OD (1982) Physiologic and anatomic investigation of a visual cortical area situated in the ventral bank of the anterior ectosylvian sulcus of the cat. Exp Brain Res 46:1–11

Neville HJ, Bavelier D, Corina D, Rauschecker JP, Karni A, Lalwani A, Braun A, Clark V, Jezzard P, Turner R (1998) Cerebral organization for language in deaf and hearing subjects: Biological constraints and effects of experience. Proc Natl Acad Sci USA 95:922–929

Nobre AC, Plunkett K (1997) The neural system of language: structure and development. Source: Curr Opin Neurobiol Unique Identifier: 97287882

Olson CR, Graybiel AM (1983) An outlying visual area in the cerebral cortex of the cat. Prog Brain Res 58:239–245

Rauschecker JP (1995) Compensatory plasticity and sensory substitution in the cerebral cortex. Trends Neurosci 18:36–43

Rauschecker JP, Aschoff A (1987) Changes in corticotectal projections of cats after visual deprivation. Neuroscience 22:S222

Rauschecker JP, Kniepert U (1994) Enhanced precision of auditory localization behavior in visually deprived cats. Eur J Neurosci 6:149–160

Rauschecker JP, Korte M (1993) Auditory compensation for early blindness in cat cerebral cortex. J Neurosci 13:4538–4548

Rauschecker JP, Tian B, Korte M (1993) Spatial tuning of single neurons in the anterior auditory areas (AAF and AEA) of the cat's cortex. Assoc Res Otolaryngol Abstr 16:109

Rice CE, Feinstein SH, Schusterman RJ (1965) Echo-detection ability of the blind: size and distance factor. J Exp Psychol 70:246–251

Tallal P, Miller SL, Bedi G, Byma G, Wang X, Nagarajan SS, Schreiner C, Jenkins WM, Merzenich MM (1996) Language comprehension in language-learning impaired children improved with acoustically modified speech. Science 271:81–84

Weeks RA, Tian B, Wessinger CM, Cohen LG, Hallett M, Rauschecker JP (1997) Identification of the inferior parietal lobule as the site of auditory space perception in humans. Neurology 48:S30.003

Wiesel TN, Hubel DH (1965) Comparison of the effects of unilateral and bilateral eye closure on cortical unit responses in kittens. J Neurophysiol 28:1029–1040

Functional Relevance of Cortical Plasticity

*L. G. Cohen, R. Chen, and P. Celnik**[*]

Summary

Different reports have demonstrated cortical plasticity associated with large hemispheric lesions, skill acquisition and use. However, it is not clear if this plasticity is always useful in terms of compensatory function. A patient who had a hemispherectomy during childhood is presented as an example of functionally relevant plasticity in the motor domain. At the time of testing, he was able to voluntarily move the affected arm. Transcranial stimulation demonstrated the development of different cortical representations for both arms in the remaining hemisphere. Therefore, the remaining hemisphere took over some of the functions of the missing hemisphere. Functional relevance of cortical plasticity can be also demonstrated when reorganization takes place across sensory modalities. Cortical areas normally processing visual information can be activated by somatosensory input. To determine if this activation contributes to behavioral performance in tactile discrimination tasks, repetitive transcranial magnetic stimulation was delivered to different scalp positions during reading of Braille and embossed Roman letters by blind and sighted subjects. Stimulation of the occipital areas disrupted reading and induced phantom tactile sensations in the early blind, but not in sighted volunteers. The occipital cortex is not only active in association with, but is one of the important functional components of the network mediating tactile discrimination in the blind. Determination of the functional role played by plasticity in each setting is important for the establishment of rational strategies for promoting recovery of function in humans.

Introduction

The human central nervous system is capable of reorganizing and adapting to new environmental requirements. There is extensive evidence for plastic changes associated with a variety of lesions like amputations (Cohen et al. 1991 a; Fuhr et al. 1992; Kew et al. 1994; Hall et al. 1990; Ridding and Rothwell 1995; Pascual-Leone et al. 1996 a; Flor et al. 1995; Knecht et al. 1995, 1996; Kew et al. 1997; Elbert

---

[*] Human Cortical Physiology Unit National Institute of Neurological Disorders and Stroke National Institutes of Health

J. Grafman / Y. Christen (Eds.)
Neuronal Plasticity:
Building a Bridge from the Laboratory to the Clinic
© Springer-Verlag Berlin Heidelberg New York 1999

et al. 1997), spinal cord injury (Cohen et al. 1991 c; Levy et al. 1990; Topka et al. 1991; Streletz et al. 1995), hemispherectomy (Cohen et al. 1991 d; Benecke et al. 1991), stroke (Turton et al. 1995, 1996; Traversa et al. 1997; Caramia et al. 1996; Bastings and Good 1997; Rapisarda et al. 1996; Catano et al. 1995, 1996, 1997; Chollet et al. 1991; Di Piero et al. 1992; Frackowiak et al. 1991; Weiller et al. 1992, 1993, 1995; Fries et al. 1991; Honda et al. 1997; Leifer et al. 1997; Weder and Seitz 1994; Hamdy et al. 1996, 1997; Netz et al. 1997), deafferentation (Brasil-Neto et al. 1992, 1993; Ridding and Rothwell 1995; Birbaumer et al. 1997; Ziemann et al. 1997), skill acquisition (Pascual-Leone et al. 1994, 1995) and use (Classen et al. 1996, 1998) in humans. However, there is not much evidence indicating that plasticity as identified using neuroimaging or neurophysiological techniques contributes to recovery or to improvement in function (Cohen et al. 1997). In this chapter, two examples of functionally relevant plasticity with a clear compensatory role will be discussed, one in the motor domain and one in the sensory domain.

Role of Ipsilateral Pathways in Motor Control

The idea that ipsilateral motor pathways play a role in recovery of motor function following stroke is not new. Fisher, for example, described two patients with good recovery from a previous stroke in whom hemiplegia reappeared following another pure motor stroke in the opposite hemisphere (Fisher 1992). A similar case was described later by Lee and van Donkelaar (1995). Noninvasive techniques have provided some evidence pointing in the same direction. A transcranial Doppler study showed a greater increase in flow velocity in the ipsilateral middle cerebral artery with movement of the recovered hand in patients with ischemic stroke when compared to the unaffected side or with control subjects (Silvestrini et al. 1995). More importantly, PET studies also suggested a role for the ipsilateral hemisphere in recovery of motor function after stroke. An initial report on patients with capsular strokes who recovered motor function showed increased cerebral blood flow (rCBF) in contralateral primary sensorimotor cortex and in the ipsilateral cerebellar hemisphere in association with finger movements of the intact hand. With movements of the affected and recovered hand, rCBF increased bilaterally in primary sensorimotor cortex, cerebellar hemispheres, insula, inferior parietal, and premotor cortices. Compared with the normal hand, the hand that recovered movement, showed increased activation of ipsilateral sensorimotor cortex, insula and inferior parietal cortex and contralateral cerebellum (Chollet et al. 1991). Similarly, increased activation of the hemisphere ipsilateral to a paretic hand was also reported by Honda et al. (1997). In a later study, Weiller et al. (1992) confirmed the earlier finding of increased activation in patients in the contralateral cerebellum, the ipsilateral insula and inferior parietal cortex, but not the ipsilateral sensorimotor cortex. Activation was greater than in normal subjects in bilateral insulae, inferior parietal, prefrontal and anterior cingulate cortices, and in the ipsilateral premotor cortex and basal ganglia. Bilateral activation of cerebral structures appeared to be one of the features associated with motor recovery from striatocapsular strokes. The individual patterns of cerebral activation in eight patients were compared with the pattern of a group of 10 normal subjects. The authors found an anterior displacement of the hand representation in the contralateral sensorimotor cortex in all patients with lesions of the posterior limb of the internal capsule. Motor pathways ipsilateral to the recovered limb were also more activated in the patients than in normal subjects, but additional activation of the ipsilateral sensorimotor cortex was only found in the four patients who exhibited associated movements of the unaffected hand with movement of the recovered hand. More recent fMRI studies have also found this activation in stroke patients in the absence of mirror or associated movements (Cramer et al. 1997; Leifer et al. 1997).

The idea that the motor cortex can play a role in motor control of the ipsilateral hand gains support from a recent neurophysiological study that studied performance of simple and complex finger sequences in normal volunteers (Chen et al. 1997). The authors studied the effects of transient disruption of activity in primary motor cortex by repetitive transcranial magnetic stimulation (TMS) on the performance of finger sequences of different complexities (Chen et al. 1997). Ten right-handed subjects were first trained to perform a simple and a complex sequence on an electronic piano with either hand. Both sequences were eight seconds long and had 16 key presses. The simple sequence was ordered and involved adjacent fingers consecutively, whereas the complex sequence was random and more difficult to execute. A water-cooled eight-shaped coil, each loop of which measures seven cm in diameter, was used. TMS at 15 Hz for 2.3 sec at an intensity of 120 % of the motor threshold was used for stimulation of the motor cortex (M1) ipsilateral to the playing hand. For M1 stimulation contralateral to the playing hand, the sequence could be disturbed at lower intensities and the stimuli were reduced to 110 % of the motor threshold. The precise timing of the key presses was recorded and the number of key press and timing errors were counted. Key press errors were defined as pressing the wrong key, pressing an extra key or omitting a key. Timing errors were defined as the time interval between key presses that were outside 2.5 standard deviations of the corresponding control interval in the same subject. The error rates were compared with those of the control condition with the stimulating coil on the scalp but directed away from the head.

As expected, contralateral M1 stimulation led to a large number of key press and timing errors with either hand. Ipsilateral M1 stimulation did not induce a significant increase in key press errors, but caused a significant increase in timing errors in both the simple and complex sequences in either hand, with higher error rates in the complex sequences than the simple sequences. With the complex sequence, the error rate was higher in the left hand than in the right. The occurrence of timing errors within the sequence was also different between the right and left sides. With the complex sequence and ipsilateral M1 stimulation, timing errors in the right hand occurred mainly during TMS whereas errors in the left hand occurred both during and slightly after the end of TMS.

While neuroimanging studies can identify the network of brain regions activated in association with performance of different tasks or stimulation, neurophysiological techniques can provide information regarding the functional role of the areas activated. From this point of view these techniques are complementary to each other. If stimulation of the motor cortex ipsilateral to a paretic hand or arm evokes ipsilateral motor responses at appropriate latencies, there might be functionally active corticomotoneuronal connections between the stimulated motor cortex and ipsilateral hand or arm muscles. If so, the stimulated hemisphere could be more likely to participate in motor control of the ipsilateral limb.

An example of this functionally relevant plasticity is the case of some patients with hemispherectomy. This surgical procedure was attempted for the first time by Lhermitte (1928) and Walter Dandy (1928) for the treatment of cerebral gliomas. In 1950, Krinauw reported 12 patients with medically intractable seizures in whom the procedure dramatically reduced seizure frequency. Since then, hemispherectomy has been used to treat pharmacologically unresponsive seizure conditions. It is a common finding that individuals with hemispherectomy performed at an early age have relatively good control of ipsilateral arm movements, particularly in the proximal segments. Stimulation of the intact hemisphere using TMS in these patients induces prominent motor responses from ipsilateral upper extremity muscles, a finding not seen under similar conditions in normal volunteers (Cohen et al. 1991 d; Benecke et al. 1991). These reports suggest that the intact hemisphere is capable of taking over certain functions of the missing hemisphere, and that this reorganization could be effective enough to allow a moderate degree of recovery in motor control. Interestingly, muscles ipsilateral to the intact hemisphere were activated by stimulation of scalp positions anterior and lateral to those activating muscles on the normal side (Fig. 1).

Further evidence pointing in this direction is that ipsilateral elbow movements were associated with rCBF increases in an area centered 1.4 cm anterior and lateral to that activated by the same movements on the normal side (Cohen et al. 1991 d). These results indicate that the intact hemisphere developed different cortical representations for the ipsilateral and contralateral arm, a form of functionally relevant plasticity. In addition to the findings reported after hemispherectomy, Carr et al. (1993; Maegaki et al. 1995) found that when the lesion takes place before birth, individual corticomotoneuronal connections in the pyramidal tract may target simultaneously ipsi- and contralateral alphamotoneuron pools. Other neurophysiological studies demonstrated that, when the stroke lesions take place later in life, ipsilateral motor responses to stimulation of one motor cortex are usually seen in the context of poor motor recovery (Turton et al. 1995, 1996; Netz et al. 1997). Therefore, a similar measure of plasticity (ipsilateral responses to stimulation of primary motor cortex) can be functionally relevant in the setting of hemispherectomy or hemispheric lesions at an early age or it can be associated with poor recovery of function in the case of lesions acquired in adulthood.

LEFT HEMISPERECTOMY

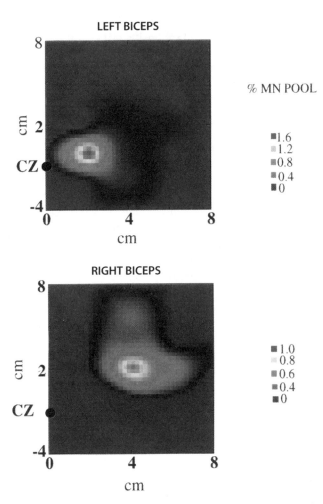

Fig. 1. Diagram representing the top of the head (right hemisphere) of a 32-year-old man who had a large congenital porencephalic cyst that had occupied most of the left hemisphere. At seven years of age he developed uncontrolled seizures and had a left hemispherectomy. The motor maps show the percentage of the alpha-motoneuron (MN) pool activated by stimulation of each scalp location in the left biceps (contralateral to the intact hemisphere) and the right biceps (ipsilateral to the intact hemisphere). Note that the remaining (right) hemisphere has a differentiated representation for both muscles. The representation for the right biceps is located more laterally and anteriorly than the representation for the left biceps. Modified from Cohen et al. 1991 d.

Cross-Modal Plasticity

The issue of plastic changes associated with visual deprivation has been present consistently in the literature for the last century. The idea that cortical areas usually dedicated to processing a specific sensory modality could change to process other modalities has fascinated neuroscientists (Pons 1996) and induced heated discussions (Rauschecker 1995; see also this volume). In the blind, for example, different studies have attempted to define if the deafferented visual cortex of individuals who lost vision at an early age can perform some functions, and if so which. While some studies in blind subjects showed that metabolic activity of the occipital cortex is unmodified in comparison to sighted controls (Phelps et al. 1981), others indicated increased occipital activation in the blind (Wanet-Defalque et al. 1988). However, these early studies failed to identify modulation of the occipital activation in association with performance of tasks like Braille reading. Electrophysiological studies also suggested some involvement of deprived occipital cortex in tactile discrimination tasks in the blind (Uhl et al. 1991). This information, together with the finding that the "mismatch negativity" in an auditory paradigm has a more posterior scalp distribution in the blind than in sighted controls and that magnetoencephalographic (MEG) potentials and brain regions activated by an auditory discrimination task are located more posteriorly in the blind (Kujala et al. 1995a, b), suggested unmasking of non-visual modalities in the visual cortex of the blind. In a similar paradigm to that used before (Uhl et al. 1991), Uhl et al. (1993) found more increased activation in SPECT of inferior occipital regions in the blind than in sighted controls, although these authors failed to find a task-dependent increase in occipital activity. A takeover by auditory and somatosensory processing of brain regions usually activated visually has been proposed in mole rats (Heil et al. 1991; Cooper et al. 1993). Blind cats appear to be more precise than sighted cats in localizing sound sources in space (Rauschecker and Kniepert 1994), as has also been reported in blind humans performing auditory localization tasks (Muchnik et al. 1991; see Rauschecker, this volume). Most of the previously described evidence (Wanet-Defalque et al. 1988; Uhl et al. 1991, 1993; Kujala et al. 1995b; Koyama et al. 1997) and a recent study using PET (Sadato et al. 1996) indicate that the occipital cortex is one of the regions activated in a distributed network associated with Braille reading and tactile discrimination tasks in blind subjects, but not in sighted volunteers (Sadato et al. 1996). In spite of this association between performance of tactile discrimination tasks and activation of the occipital cortex, the role of this occipital activation remained elusive. We therefore formulated the question whether the occipital cortex processes this information in a functionally relevant way in the sense of sensory compensation. To address this issue, we studied a group of subjects who became blind at an early age and a group of sighted controls. Both invasive (Penfield and Roberts 1959; Ojemann 1983) and noninvasive (Cohen et al. 1991b; Pascual-Leone et al. 1996b) cortical stimulation can transiently disrupt cognitive functions (e. g., speech). Since trains of stimuli are more effective than single stimuli in inducing these effects, we used this repetitive TMS instead of single pulse TMS (Luders et al. 1987; Henderson et al. 1979; Brindley and Lewin 1968). This task disruption by focal stimulation has been interpreted as a sign that the stimulated region is functionally significant for performance (Ojemann 1983). Noninvasive TMS (Barker et al. 1985) applied to occipital regions in the sighted can transiently suppress visual perception of letters (Amassian et al. 1989) and also extrafoveal targets (Epstein and Zangaladze 1996). This effect probably takes place by interference with visual calcarine (Amassian et al. 1989) and association cortical areas at depths of 1.5 to 2.25 cm below the scalp surface (Epstein et al. 1996). Therefore, TMS is an appropriate technique to induce transient disruption of function during Braille reading (Fig. 2a).

We used TMS to disrupt the function of different cortical areas transiently (Chen et al. 1997; Gerloff et al. 1997) during identification of Braille and embossed Roman letters (Cohen et al. 1997). The parameters of TMS used in this protocol were approved by the NINDS Investigational Review Board under a special Investigational Device Exception from the Food and Drug Administration and are within our most recent safety standards published by Chen et al. (in press) and presently in use at the National Institutes of Health. TMS was delivered at 10 Hz for 3 sec. with a stimulus intensity of 10 % above motor threshold for a muscle involved in the reading task (first dorsal interosseus) at rest.

Five subjects who became blind early in life and who were experienced Braille readers were studied while reading strings of "grade I" non-contracted, non-word Braille letters, and five sighted volunteers and four of the early blind subjects were also studied while performing a tactile discrimination task requiring identification of the same embossed Roman letters. Letters were presented with a specially designed device (Fig. 2b). Subjects were asked to identify and read aloud letter by letter as quickly and accurately as possible. Overall accuracy in reading performance before TMS was similar in the different groups. Midoccipital stimulation induced more errors than the control condition (stimulation in the air; Fig. 2c). In addition, stimulation of occipital positions occasionally elicited distorted somatosensory perceptions. Blind subjects reported negative ("missing dots"), positive ("phantom dots"), and confusing sensations ("dots don't make sense"). Blind and sighted subjects performing the same task (reading embossed Roman letters) showed a different effect during TMS. Midoccipital stimulation induced more errors than control (stimulation in the air) in the blind but not in sighted volunteers, supporting the view that the occipital cortex is functionally active in spite of decades of visual deafferentation (Rushton and Brindley 1978; Schmidt et al. 1996), and that it is engaged in active and meaningful processing of tactile information related to Braille reading and other tactile discrimination tasks. Since sensory processing for touch and vision appear to be segregated up to their arrival in primary reception areas, the earliest convergence of visual and somatosensory information in sighted mammals occurs at cortical association sites (Pons 1996). One possible explanation for our findings is that connections between parietal and visual association areas mediate the transfer of somatosensory information to the occipital cortex in blind subjects (Bruce et al. 1981). But what is the kind of perception that a blind subject experiences when a

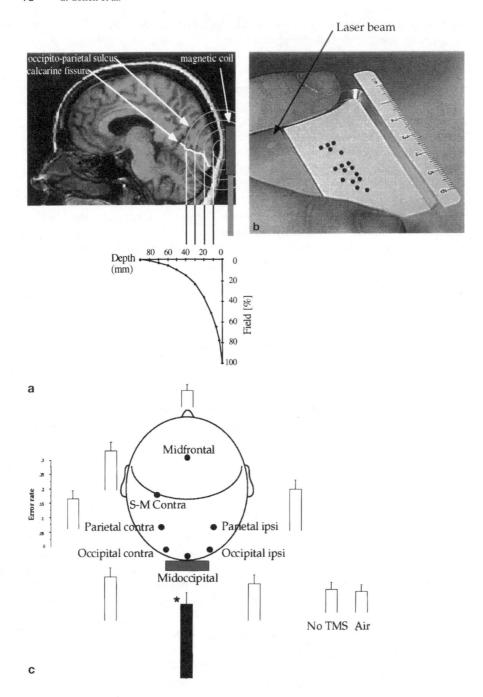

◄ **Fig. 2. a.** T1-weighted conventional MRI (sagittal slice, 1.5 Tesla) of a sighted individual. The magnetic coil is shown positioned at the midoccipital site. Concentric electric field lines are displayed at 1 cm distances from the coil center. The graph shows the marked drop in the magnitude of the TMS-induced electric field as a function of the distance from the coil center (as calculated for a circular coil; Roth et al. 1991). Therefore, the TMS-induced electric field is largest in superficial cortical regions (see also Epstein et al. 1996). **b.** Reading window where Braille letters or embossed Roman letters were presented. TMS trains of 3 sec were triggered when the reading finger crossed the laser beam at the beginning of the reading task. **c.** Diagram of the top of the head. Bars show the error rates induced by stimulation of different scalp locations in the early blind. Note that stimulation of the midoccipital location induced the largest number of errors. S-M, sensorimotor cortex; contra, contralateral; ipsi, ipsilateral. Modified from Cohen et al. 1997.

visual area is activated by a somatosensory stimulus? It is possible that the additional activation of occipital cortex in the blind adds certain characteristics to the somatosensory perceptions involved in Braille reading. The occasional induction of complex sensations (phantom dots or extra dots) with occipital TMS supports this interpretation. Therefore, the effects of midoccipital TMS were likely related to interference with complex discriminative operations performed by the occipital cortex in the early blind.

In the sighted subjects, stimulation of the occipital cortex did not affect identification of embossed Roman letters or induce abnormal somatosensory perceptions. In combination with the decrease of occipital activity on PET in subjects performing a similar task (Sadato et al. 1996), these results suggest that sighted individuals do not normally utilize the occipital cortex for identification of embossed Roman letters as the blind do for Braille and Roman letter reading.

The finding that the occipital cortex is a critical functioning component of the network involved in Braille reading supports the idea that perceptions are dynamically determined by the characteristics of the sensory inputs rather than only by the brain region that receives those inputs, at least in the case of early blindness (Pons 1996; Rauschecker 1995). These results indicate that cross-modal plasticity, as identified electrophysiologically or by neuroimaging techniques in humans, may play a functionally compensatory role.

References

Amassian VE, Cracco RQ, Maccabee PJ, Cracco JB, Rudell A, Eberle L (1989) Suppression of visual perception by magnetic coil stimulation of human occipital cortex. Electroencephalogr Clin Neurophysiol 74:458–462

Barker AT, Jalinous R, Freeston IL (1985) Noninvasive magnetic stimulation of human motor cortex. Lancet 1:1106–1107

Bastings EP, Good DC (1997) Changes in motor cortical representations after stroke: correlations between clinical observations and magnetic stimulation mapping studies. Neurology 48 (Suppl 2):A 414

Benecke R, Meyer BU, Freund HJ (1991) Reorganisation of descending motor pathways in patients after hemispherectomy and severe hemispheric lesions demonstrated by magnetic brain stimulation. Exp Brain Res 83:419–426

Birbaumer N, Lutzenberger W, Montoya P, Larbig W, Unertl K, Topfner S, Grodd W, Taub E, Flor H (1997) Effects of regional anesthesia in phantom limb pain are mirrored in changes in cortical reorganization. J Neurosci 17:5503–5508

Brasil-Neto JP, Cohen LG, Pascual-Leone A, Jabir FK, Wall RT, Hallett M (1992) Rapid reversible modulation of human motor outputs after transient deafferentation of the forearm: a study with transcranial magnetic stimulation. Neurology 42:1302–1306

Brasil-Neto JP, Valls Solé J, Pascual-Leone A, Cammarota A, Amassian VE, Cracco R, Maccabee P, Cracco J, Hallett M, Cohen LG (1993) Rapid modulation of human cortical motor outputs following ischemic nerve block. Brain 116:511–525

Brindley GS, Lewin WS (1968) The sensations produced by electrical stimulation of the visual cortex. J Physiol (Lond) 196:479–493

Bruce C, Desimone R, Gross CG (1981) Visual properties of neurons in a polysensory area in superior temporal sulcus of the macaque. J Neurophysiol 46:369–384

Caramia MD, Iani C, Bernardi G (1996) Cerebral plasticity after stroke as revealed by ipsilateral responses to magnetic stimulation. Neuroreport 7:1756–1760

Carr LJ, Harrison LM, Evans AL, Stephens JA (1993) Patterns of central motor reorganization in hemiplegic cerebral palsy. Brain 116:1233–1247

Catano A, Houa M, Caroyer JM, Ducarne H, Noel P (1995) Magnetic transcranial stimulation in non-haemorrhagic sylvian strokes: interest of facilitation for early functional prognosis. Electroencephalogr Clin Neurophysiol 97:349–354

Catano A, Houa M, Caroyer JM, Ducarne H, Noel P (1996) Magnetic transcranial stimulation in acute stroke: early excitation threshold and functional prognosis. Electroencephalogr Clin Neurophysiol 101:233–239

Catano A, Houa M, Noel P (1997) Magnetic transcranial stimulation: dissociation of excitatory and inhibitory mechanisms in acute strokes. Electroencephalogr Clin Neurophysiol 105:29–36

Chen R, Gerloff C, Hallett M, Cohen LG (1997) Involvement of the ipsilateral motor cortex in finger movements of different complexities. Ann Neurol 41:247–254

Chen R, Gerloff C, Classen J, Wassermann EM, Hallett M, Cohen LG (1998) Safety of different inter-train intervals for repetitive magnetic stimulation and recommendations for safe ranges of stimulation parameters. Electroencephalogr Clin Neurophysiol, in press

Chollet F, DiPiero V, Wise RJ, Brooks DJ, Dolan RJ, Frackowiak RS (1991) The functional anatomy of motor recovery after stroke in humans: a study with positron emission tomography. Ann Neurol 29:63–71

Classen J, Liepert A, Hallett M, Cohen LG (1996) Use-dependent modulation of movement representation in the human motor cortex. Soc Neurosci Abstr 22:1452

Classen J, Liepert A, Wise SP, Hallett M, Cohen LG (1998) Rapid plasticity of human cortical movement representation induced by practice. J Neurophysiol, in press

Cohen LG, Bandinelli S, Findlay TW, Hallett M (1991 a) Motor reorganization after upper limb amputation in man. Brain 114:615–627

Cohen LG, Bandinelli S, Sato S, Kufta C, Hallett M (1991 b) Attenuation in detection of somatosensory stimuli by transcranial magnetic stimulation. Electroencephalogr Clin Neurophysiol 81:366–376

Cohen LG, Topka H, Cole RA, Hallett M (1991 c) Leg paresthesias induced by magnetic brain stimulation in patients with thoracic spinal cord injury. Neurology 41:1283–1288

Cohen LG, Zeffiro T, Bookheimer S, Wassermann EM, Fuhr P, Matsumoto J, Toro C, Hallett M (1991 d) Reorganization in motor pathways following a large congenital hemispheric lesion: different motor representation areas for ipsi- and contralateral muscles. Physiol 438:33

Cohen LG, Celnik P, Pascual-Leone A, Corwell B, Faiz L, Honda M, Dambrosia J, Sadato N, Hallett M (1997) Functional relevance of cross-modal plasticity in the blind. Nature 389:180–183

Cooper HM, Herbin M, Nevo E (1993) Visual system of a naturally microphthalmic mammal: the blind mole rat, Spalax ehrenbergi. J Comp Neurol 328:313–350

Cramer SC, Nelles G, Benson RR, Kaplan JD, Parker RA, Kwong KK, Kennedy DN, Finklestein SP, Rosen BR (1997) Simultaneous measurement of cerebral blood flow and functional MRI signal in the evaluation of stroke recovery mechanism. Neurology 48 (Suppl 2): A 415

Dandy W (1928) Removal of right cerebral hemisphere for certain tumors with hemiplegia: preliminary report. JAMA 90:823–825

Di Piero V, Chollet FM, MacCarthy P, Lenzi GL, Frackowiack RSJ (1992) Motor recovery after acute ischaemic stroke: a metabolic study. J Neurol Neurosurg Psych 55:990–996

Elbert T, Sterr A, Flor H, Rockstroh B, Knetch S, Pantev C, Wienbruch C, Taub E (1997) Input-increase and input-decrease types of cortical reorganization after upper extremity amputation in humans. Exp Brain Res 117:161–164

Epstein CM, Zangaladze A (1996) Magnetic coil suppression of extrafoveal visual perception using disappearance targets. J Clin Neurophysiol 13:242–246

Epstein CM, Verson R, Zangaladze A (1996) Magnetic coil suppression of visual perception at an extracalcarine site. J Clin Neurophysiol 13:247–252

Fisher CM (1992) Concerning the mechanism of recovery in stroke hemiplegia. Can J Neurolog Sci 19:57–63

Flor H, Elbert T, Knecht S, Wienbruch C, Pantev C, Birbaumer N, Larbig W, Taub E (1995) Phantom limb pain as a perceptual correlate of cortical reorganization following arm amputation. Nature 375:482–484

Frackowiak RS, Weiller C, Chollet F (1991) The functional anatomy of recovery from brain injury. Ciba Found Symp 163:235–244

Fries W, Danek A, Witt TN (1991) Motor responses after a transcranial electrical stimulation of cerebral hemispheres with a degenerated corticospinal tract. Ann Neurol 29:646–650

Fuhr P, Cohen LG, Dang N, Findley TW, Haghighi S, Oro J, Hallett M (1992) Physiological analysis of motor reorganization following lower limb amputation. Electroencephalogr Clin Neurophysiol 85:53–60

Gerloff C, Corwell B, Chen R, Hallett M, Cohen LG (1997) Stimulation over the human supplementary motor area interferes with the organization of future elements in complex motor sequences. Brain 120:1587–1602

Hall EJ, Flament D, Fraser C, Lemon RN (1990) Non-invasive brain stimulation reveals reorganized cortical outputs in amputees. Neurosci Lett 116:379–386

Hamdy S, Aziz Q, Rothwell JC, Singh KD, Barlow J, Hughes DG, Tallis RC, Thompson DG (1996) The cortical topography of human swallowing musculature in health and disease [see comments]. Nature Med 2:1217–1224

Hamdy S, Aziz Q, Rothwell JC, Crone R, Hughes D, Tallis RC, Thompson DG (1997) Explaining oropharyngeal dysphagia after unilateral hemispheric stroke. Lancet 350:686–692

Heil P, Bronchti G, Wollberg Z, Scheich H (1991) Invasion of visual cortex by the auditory system in the naturally blind mole rat. Neuroreport 2:735–738

Henderson DC, Evans JR, Dobelle WH (1979) The relationship between stimulus parameters and phosphene/brightness during stimulation of human visual cortex. Trans Am Soc Artif Intern Organs 25:367–371

Honda M, Nagamine T, Fukuyama H, Yonekura Y, Kimura J, Shibasaki H (1997) Movement-related cortical potentials and regional cerebral blood flow change in patients with stroke after motor recovery. J Neurol Sci 146:117–126

Kew JJ, Ridding MC, Rothwell JC, Passingham RE, Leigh PN, Sooriakumaran S, Frackowiak RS, Brooks DJ (1994) Reorganization of cortical blood flow and transcranial magnetic stimulation maps in human subjects after upper limb amputation. J Neurophysiol 72:2517–2524

Kew JJ, Halligan PW, Marshall JC, Passingham RE, Rothwell JC, Ridding MC, Marsden CD, Brooks DJ (1997) Abnormal access of axial vibrotactile input to deafferented somatosensory cortex in human upper limb amputees. J Neurophysiol 77:2753–2764

Knecht S, Henningsen H, Elbert T, Flor H, Hohling C, Pantev C, Birbaumer N, Taub E (1995) Cortical reorganization in human amputes and mislocalization of painful stimuli to the phantom limb. Neurosci Lett 201:262–264

Knecht S, Henningsen H, Elbert T, Flor H, Hohling C, Pantev C, Taub E (1996) Reorganization and perceptual changes after amputation. Brain 119:1213–1219

Koyama K, Gerloff C, Celnik P, Cohen LG, Classen J, Honda M, Hallett M (1997) Functional cooperativity of visual, motor, and premotor areas during Braille reading in patients suffering from peripheral blindness early in life. Neurology 48 (Suppl 2):A 305

Krinauw RA (1950) Infantile hemiplegia treated by removing one cerebral hemisphere. J Neurol Neurosurg Psych 13:243–267

Kujala T, Alho K, Kekoni J, Hamalainen H, Reinikainen K, Salonen O, Standertskjold NC, Naatanen R (1995a) Auditory and somatosensory event-related brain potentials in early blind humans. Exp Brain Res 104:519–526

Kujala T, Huotilainen M, Sinkkonen J, Ahonen AI, Alho K, Hamalainen MS, Ilmoniemi RJ, Kajola M, Knuutila JET, Lavikainen J, Salonen O, Simola J, Sandertskjold-Nordenstam C-G, Tiitinen H, Tissari SO, Naatanen R (1995b) Visual cortex activation in blind subjects during sound discrimination. Neurosci Lett 183:143–146

Lee RG, van Donkelaar P (1995) Mechanisms underlying functional recovery following stroke. Can J Neurol Sci 22:257–263

Leifer D, Zhong J, Fulbright RK, Graham GD, Prichard JW, Gore JC (1997) Functional MRI reveals changes in brain activation during motor tasks by stroke patients. Neurology 48 (Suppl 2):A 415

Levy WJ, Amassian VE, Traad M, Cadwell J (1990) Focal magnetic coil stimulation reveals motor cortical system reorganized in humans after traumatic quadriplegia. Brain Res 510:130–134

Lhermitte J (1928) L'ablation complète de l'hemisphere droit dans les cas de tumeur cérébrale localisée compliquée d'hémiplégie: la décérebration suprathalamique unilaterale chez l'homme. Encephale 23:314–323

Luders H, Lesser RP, Dinner DS, Morris HH, Hahn JF, Friedman L, Skipper G, Wyllie E, Friedman D (1987) Commentary: Chronic intracranial recording and stimulation with subdural electrodes. In: J. Engel, Ed. Surgical treatment of the epilepsies. New York; Raven Press, 297–321

Maegaki Y, Maeoka Y, Takeshita K (1995) Plasticity of central motor pathways in hemiplegic children with large hemispheric lesions. Electroencephalogr Clin Neurophysiol 97:S 192

Muchnik C, Efrati M, Nemeth E, Malin M, Hildesheimer M (1991) Central auditory skills in blind and sighted subjects. Scand Audiol 20:19–23

Netz J, Lammers T, Homberg V (1997) Reorganization of motor output in the non-affected hemisphere after stroke. Brain 120:1579–1586

Ojemann GA (1983) Brain organization for language from the perspective of electrical stimulation mapping. Behav Brain Sci 6:190–206

Pascual-Leone A, Grafman J, Hallett M (1994) Modulation of cortical motor output maps during development of implicit and explicit knowledge [see comments]. Science 263:1287–1289

Pascual-Leone A, Nguyet D, Cohen LG, Brasil NJ, Cammarota A, Hallett M (1995) Modulation of muscle responses evoked by transcranial magnetic stimulation during the acquisition of new fine motor skills. J Neurophysiol 74:1037–1045

Pascual-Leone A, Peris M, Tormos JM, Pascual AP, Catala MD (1996a) Reorganization of human cortical motor output maps following traumatic forearm amputation. Neuroreport 7:2068–2070

Pascual-Leone A, Wassermann E, Grafman J, Hallett M (1996b) The role of dorsolateral prefrontal cortex in implicit procedural learning. Exp Brain Res 107:479–485

Penfield W, Roberts L (1959) Speech and brain mechanisms. Princeton, NJ, Princeton University Press

Phelps ME, Mazziotta JC, Kuhl DE, Newer M, Packwood J, Metter J, Engel J (1981) Tomographic mapping of human cerebral metabolism: visual stimulation and deprivation. Neurology 31:517–529

Pons T (1996) Novel sensations in the congenitally blind. Nature 380:479–480

Rapisarda G, Bastings E, de Noordhout AM, Pennisi G, Delwaide PJ (1996) Can motor recovery in stroke patients be predicted by early transcranial magnetic stimulation? Stroke 27:2191–2196

Rauschecker JP (1995) Compensatory plasticity and sensory substitution in the cerebral cortex. Trends Neurosci 18:36–43

Rauschecker JP, Kniepert U (1994) Auditory localization behaviour in visually deprived cats. Eur J Neurosci 6:149–160

Ridding MC, Rothwell JC (1995) Reorganization in human motor cortex. Can J Physiol Pharmacol 73:218–222

Roth BJ, Saypol JM, Hallett M, Cohen LG (1991) A theoretical calculation of the electric field induced in the cortex during magnetic stimulation. Electroencephalogr Clin Neurophysiol 81:47–56

Rushton DN, Brindley GS (1978) Properties of cortical electrical phosphenes. In: S. J. Cool and E. L. Smith, Ed. Frontiers in visual science. New York Springer-Verlag, 574–593

Sadato N, Pascual-Leone A, Grafman J, Ibañez V, Deiber M-P, Dold G, Hallet M (1996) Activation of the primary visual cortex by Braille reading in blind subjects. Nature 380:526–528

Schmidt EM, Bak MJ, Hambrecht FT, Kufta CV, O'Rourke DK, Vallabhanath P (1996) Feasibility of a visual prosthesis for the blind based on intracortical microstimulation of the visual cortex. Brain 119:507–522

Silvestrini M, Troisi E, Matteis M, Cupini LM, Caltagirone C (1995) Involvement of the healthy hemisphere in recovery from aphasia and motor deficit in patients with cortical ischemic infarction: a transcranial Doppler study. Neurology 45:1815–1820

Streletz LJ, Belevich JK, Jones SM, Bhushan A, Shah SH, Herbison GJ (1995) Transcranial magnetic stimulation: cortical motor maps in acute spinal cord injury. Brain Topogr 7:245–250

Topka H, Cohen LG, Cole RA, Hallett M (1991) Reorganization of corticospinal pathways following spinal cord injury. Neurology 41:1276–1283

Traversa R, Cicinelli P, Bassi A, Rossini PM, Bernardi G (1997) Mapping of motor cortical reorganization after stroke. A brain stimulation study with focal magnetic pulses. Stroke 28:110–117

Turton A, Wroe S, Trepti N, Fraser C, Lemon RN (1995) Ipsilateral EMG responses to transcranial magnetic stimulation during recovery of arm and hand function after stroke. Electroencephalogr Clin Neurophysiol 97:S 192

Turton A, Wroe S, Trepte N, Fraser C, Lemon RN (1996) Contralateral and ipsilaterel EMG responses to transcranial magnetic during recovery of arm hand function after stroke. Electroencephalogr Clin Neurophysiol 101:316–328

Uhl F, Franzen P, Lindinger G, Lang W, Deecke L (1991) On the functionality of the visually deprived occipital cortex in early blind persons. Neurosci Lett 124:256–259

Uhl F, Franzen P, Podreka I, Steiner M, Deecke L (1993) Increased regional cerebral blood flow in inferior occipital cortex and cerebellum of early blind humans. Neurosci Lett 150:162–164

Wanet-Defalque M-C, Veraart C, De Volder A, Metz R, Michel C, Dooms G, Goffinet A (1988) High metabolic activity in the visual cortex of early blind subjects. Brain Res 446:369–373

Weder B, Seitz RJ (1994) Deficient cerebral activation pattern in stroke recovery. Neuroreport 5:457–460

Weiller C, Chollet F, Friston KJ, Wise RJ, Frackowiak RS (1992) Functional reorganization of the brain in recovery from striatocapsular infarction in man. Ann Neurol 31:463–472

Weiller C, Ramsay SC, Wise RJ, Friston KJ, Frackowiak RS (1993) Individual patterns of functional reorganization in the human cerebral cortex after capsular infarction. Ann Neurol 33:181–189

Weiller C, Isensee C, Rijntjes M, Huber W, Muller S, Bier D, Dutschka K, Woods RP, Noth J, Diener HC (1995) Recovery from Wernicke's aphasia: a positron emission tomographic study. Ann Neurol 37:723–732

Ziemann U, Corwell B, Hallett M, Cohen LG (1997) Modulation of plastic changes in human motor cortex after forearm ischemic nerve block. Soc Neurosci Abs 2237

"Anomalous" Representations and Perceptions: Implications for Human Neuroplasticity

S. Aglioti *

Summary

Changes in the perception and representation of the body observed in brain damaged and amputee patients hint at the plastic nature of the body schema. Evidence for integration of external objects into the body schema comes from a woman with a large right-hemisphere stroke who affirmed that the paralysed left hand was not her own but belonged to someone else. Although able to see and describe the rings she had worn for years and was currently wearing on her left, now disowned hand, this patient resolutely denied their ownership. By contrast, she had no difficulties in recognising these rings as her own when they were shifted to her right hand, or displayed by the examiner in front of her. Similarly, she promptly acknowledged ownership of other personal belongings that in her previous experience had not been ordinarily associated with the left hand (e.g., a comb). Complex dynamic aspects of the body schema are also revealed by the recent evidence in limb or breast amputees that vivid phantom sensations can arise as a result of tactile stimulations applied to body regions distant from the amputation line. Sensations in the phantom hand, for example, can be elicited by tactile stimuli delivered to the lower face on the side of the amputation. Like the concurrent veridical facial sensations, the evoked phantom sensations may convey precise information about different features of the facial stimuli. Given the representional contiguity of face and hand, phantom hand sensations from facial stimulation are probably caused by an appropriation of the original somatosensory representation of the lost hand by sensory inputs inherent to the adjacent face representation.

The representation of the body, however, has some aspects of stability. Observations in amputees indicate, for example, that phantom perceptions may persist over decades, thus suggesting that parts of the brain may be quasi-permanently committed to the representation of a given body part. In the same vein, stimulation of the somatosensory cortex in amputees who do not feel any phantom perceptions may resurrect the phantom limb even 25 years after the amputation.

* Dipartimento di Scienze Neurologiche e della Visione, Sezione di Fisiologia Umana, Università' di Verona, Italy

J. Grafman / Y. Christen (Eds.)
Neuronal Plasticity:
Building a Bridge from the Laboratory to the Clinic
© Springer-Verlag Berlin Heidelberg New York 1999

The Concept of body Schema

The term body schema (or body image) alludes to the complex of sense-impressions, perceptions, and ideas about the dynamic organisation of one's own body. This psychophysiological construct is built-up and maintained because of the interaction between a distributed neural network in the central nervous system and dynamic inputs from tactile, proprioceptive, vestibular, and visual periphery (Critchley 1979; Frederiks 1985a; Melzack 1990). Physiological and pathological perturbations, such as learning or brain lesions or sensory deprivation, can induce changes in the way the human body or parts of it are perceived and represented. Thus, the issue discussed in the present chapter is relevant for cognitive neuroscience and neuropsychiatry.

The terms body schema and body image are often used interchangeably. The latter term, however, is more commonly employed in the psychiatric literature to refer to the psychocultural components of the concept. Body-centred delusions such as underestimation of the size of bodily parts are often observed in major psychiatric illnesses like schizophrenia, where such symptoms are more frequently related to the left side, and depression, where they are more frequently related to the right side. Also hypochondriacs tend to refer more frequently to the right side of the body when expressing their complaints (e.g., a pain in an arm). These side differences may be pathologic expressions of the asymmetric functioning of the cerebral hemispheres (McGilchrist and Cutting 1995). Body-related disorders dominate the clinical picture in the depersonalisation syndrome, characterised by a persistent feeling of living outside one's own body, and in dysmorphophobia or body dysmorphic disorder (BDD), a condition first identified by the Italian psychiatrist, Enrico Morselli (Morselli 1886). Patients with BDD are morbidly preoccupied with real or imaginary physical flaws, concerning, for example, the shape of the nose, the hair appearance, the size of the penis or breast, etc., to the point of seeking inappropriate and unnecessary surgical corrections or even suffering from self-inflicted injuries. Patients with Cotard's syndrome are affected by a nihilistic delusion about their body that suggests a specific disorder of the neural bases of corporeal and/or egocentric space awareness (Young et al. 1994). All these complex psychiatric conditions can now be subjected to quasi-experimental analyses using modern methods for functional brain imaging, so that advances in the understanding of the neural bases of psychiatric alterations of the body schema and body image are foreseeable in the future.

Changes in Perception and Representation of the Human Body Following Brain Lesions

Misperceptions and misrepresentations of the body consequent to focal brain lesions may reflect a selective damage of neural loci dedicated to the whole body or parts of it. According to Melzack (1990, 1992), awareness of the body relies upon a large neural network where the somatosensory cortex, the posterior parietal lobe and the insular cortex play crucial and different roles. Lesions of the primary somatosensory cortex obviously modify feelings about the body, insofar as they induce deficits in the tactile and proprioceptive modalities, but there is no evidence that they may bring about alterations of higher-order body awareness. On the other hand, it is not uncommon that patients with lesions centered upon the posterior parietal lobe, particularly on the right side, appear anosognosic for their hemiplegia and their somatosensory defects, such that they minimise their disturbances and sometimes vehemently deny being impaired at all (McGlynn and Schacter 1989; Levine et al. 1991). Feelings of non-belonging or denial of the ownership of a body part, most often the upper limb, and even misoplegia, i.e., hatred of the hemiparetic limbs, have been reported in other brain damaged patients (Bisiach et al. 1991; Rode et al. 1992; Halligan et al. 1995; Moss and Turnbull 1996). The neglected or disowned body parts appear to be constantly expunged from the patients' mental representation of their own body, so that in order to account for the material existence of these parts they usually resort to improbable rationalisations and confabulations – somatoparaphrenia (Bisiach and Geminiani 1991; Ramachandran 1995). In keeping with lesion studies is a recent PET finding of a specific posterior parietal system (superior parietal cortex, the intraparietal sulcus, and the adjacent rostralmost part of the inferior parietal lobule) involved in mental transformations of the body in space (Bonda et al. 1995). There is evidence to suggest the involvement of insular structures in the representation of the body, since lesions in this region can cause somatic hallucinations (Roper et al. 1993), and electrical stimulation near the insula induces illusions of changes in body position and feelings of being outside one's body (Penfield and Jasper 1954). Insular structures have probably to do primarily with the emotional aspects of body awareness.

The possibility that the body schema may be extended to include inanimate objects bearing systematic relations to the body emphasizes the plasticity of this construct. It is an intuitive concept that the body schema can be extended to include non-corporeal objects, such as clothes and tools, for example the feather in the hat of Edwardian women, the knife in the surgeon's hand, the white stick in the image of a blind man (Critchley 1979; Head and Holmes 1911). No experimental support for this speculation had been provided before the description of a 73-year-old woman with a large right-hemisphere stroke who denied resolutely the ownership of her left hand (Aglioti et al. 1996). Amazingly, the disavowal reaction extended from the hand to objects perceptually and representationally related to the deluded body part. Indeed, the patient denied the ownership of her rings and watch when they were worn on the left hand. By contrast, the same objects were correctly recognized as her own when worn on the right, normal hand (Fig. 1).

This effect was specific to objects related to the denied body part because other personal objects (e.g., pins, earrings, comb) were always recognized as her own (Table 1).

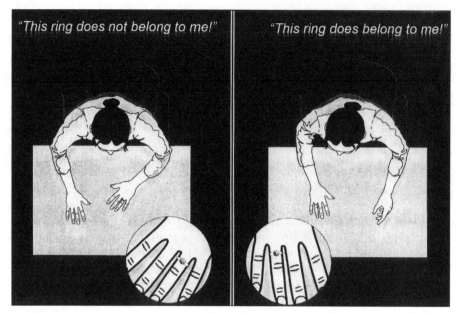

Fig. 1. Schematic representation of the patient's behaviour contingent on the position of the ring on the hands. The lady-of-the-rings was tested while lying in a bed. The sketchy drawing shows a from-above view for the sake of clarity.

Table 1. Objects chosen for testing were assigned to one of four possible categories defined according to two properties relevant to the task, namely ownership and perceputo-representational relationships with the deluded body part. Legend: B: objects belonging to the patient; NB: objects non belonging to the patient; R: objects related to the deluded body parts; NR: objects non related to the deluded body parts. 1: B-R (golden ring, garnet ring, brown strap watch); 2: B-NR (blue comb, knife, clock, holy icons, earrings); 3: NB-R (lapis lazuli ring, diamond ring, black strap watch); 4: NB-NR (key ring, scissors, padlock).

Hand	Object category	Date of testing		
		8/11/95	9/11/95	10/11/95
Left	1	0/6	0/6	0/6
	2	4/4	4/4	2/2
	3	2/2	6/6	5/5
	4	4/4	4/4	4/4
Right	1	4/4	6/6	6/6
	2	4/4	2/2	2/2
	3	2/2	4/4	4/4
	4	4/4	4/4	2/2

Moreover, it appeared that the ability of this patient to retrieve information about the rings was very limited when the rings were worn on the deluded body part. By contrast, a lot of information on the history of each particular ring was retrieved when it was on the right hand. The impairment in recognising as her own objects related to the deluded body part strictly paralleled the somatoparaphrenic disturbance; such impairment was, in fact, mild when the delusion symptom started to clear up and absent when delusion was no longer present.

Evidence for an extended body schema has also been obtained by recording, from the caudal postcentral gyrus of the monkey cortex, the activity of neurons that respond to somatosensory and visual stimuli arising from the hands. If the monkey retrieves food with one hand, the visual receptive fields of these neurons are limited to the performing hand, but if the retrieval is done with the help of a rake, the visual receptive fields expand to include both the hand and the tool (Iriki et al. 1996).

Changes in the Perception and Representation of the Body Following Peripheral Perturbations

Somatosensory inputs to the brain, especially from proprioceptors, are essential for bodily awareness. This is suggested by the striking illusory change in the perceived position and shape of the body induced in normal subjects by an intense proprioceptive stimulation. Lackner (1988), for example, delivered muscle vibration to the biceps brachii in blindfolded normal subjects who kept their index and middle fingers in contact with their nose. Although the stimulation did not induce any overt postural changes, 10 of 14 subjects reported an illusory extension of the forearm; remarkably, five subjects also reported a very puzzling elongation of the nose, by as much as 30 cm, that appeared to follow the positional changes of index and middle fingers. This result hints at the highly plastic nature of the body schema by suggesting that information from the sensory periphery can modify on-line the way in which the body is perceived.

Phantom phenomena, i.e., sensations and perceptions originating from a removed body part, like upper or lower limbs, breast, jaw, penis and so on, have been repeatedly reported (Cronholm 1951; Frederiks 1985 a, b; Melzack 1992). The analysis of "phantom" sensations felt in a no longer existing bodily structure provides important insights into the mechanisms of corporeal awareness. Visual, auditory, and olfactory phantom sensations have also been reported after partial or complete deafferentation of the corresponding sense organs (Melzack 1992); there is no doubt, however, that the most obvious phantom phenomena are somaesthetic in nature, with all submodalities, from pain to feeling of movement, from touch to thermal feelings, being represented in the phantom experience. Peripheral activation of sensory nerves at the amputation scar can of course contribute to such an experience, but it is clear evidence that phantom phenomena have relevant central components. Melzack (1992) postulated that the body schema is subserved by a specific distributed neural network or neuromatrix, largely prewired by genetics but open to the continuous shaping influence of experience, which includes the somatosensory system, reticular afferents to the limbic system, and cortical regions important for both self-recognition and the recognition of external objects and entities. Phantom phenomena would be primarily caused by the persisting activity of components of the neuromatrix that have been deprived of their normal inputs because of the loss of a body part, and by the brain's interpretation of this activity as originating from the lost part. The relevance of central components in determining phantom perceptions is also suggested by their disappearance after lesions to the right posterior parietal lobe (Appenzeller and Bicknell 1969; Melzack 1992; Aglioti et al. 1994a). Moreover, phantom perceptions are most frequent and vivid following limb amputation, probably due to the functional relevance and the extensive cerebral representation of these body parts. The phantom limb may be perceived as identical in shape to the former real limb, thus suggesting that structures in the central nervous system are committed to the representation of that body part. As time elapses, however, the phantom limb may shrink in such a way that the hand is perceived as attached directly to the shoulder (telescoping phenomena, Katz 1992; Katz and Melzack 1990). This effect may reflect reorganizational changes in the neural substrates formerly representing the lost body part (Katz 1992).

Psychophysical testing in amputee patients who experienced phantom perceptions provides novel information on possible perceptual correlates of the plastic changes observed in animal research (Pons et al. 1991; Kaas 1991; Merzenich and Sameshima 1993; Weinberger 1995).

The simple experimental paradigm used in this type of research is described elsewhere (Ramachandran et al. 1992; Aglioti et al. 1994a, b, c). Thus, only a short report is provided below. In a preliminary interview, the patients were asked about any "anomalous" sensations that may have occurred after the amputation (how long they have lasted, when they started and whether or not these sensations were still present at the testing period). During the testing sessions, patients were required to report on the sensations elicited by somatic stimuli delivered on several skin points both ipsilateral and contralateral to the amputation. In some of the patients who had undergone the amputation of an upper limb or hand, the stimulation of the stump and lower face (Ramachandran et al. 1992; Halligan et al. 1993) or the neck ipsilateral to the amputation (Aglioti et al. 1994a) elicited sensations on the phantom in addition to a local sensation. In patients with lower limb amputations, phantom sensations were evoked by stimuli on the stump, even remote from the amputation line; similar double sensations were reported during sexual intercourse and defecation (Aglioti et al. 1994b). Evoked phantom sensations may present the same perceptual features (form, numerosity, somatic sub-modality) of the local sensations. Finally, in some mastectomy patients, phantom nipple sensations were evoked by stimuli on the trunk and the pinna ipsilateral to the breast amputation (Aglioti et al. 1994c). A schematic representation of the pattern of evoked sensations in representative subjects in provided in Figure 2.

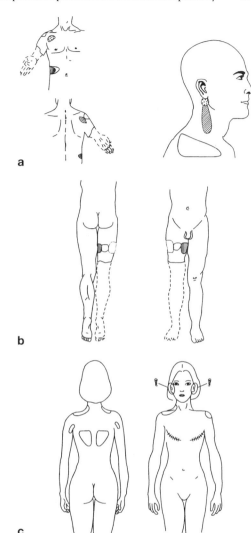

Fig. 2. Pattern of skin regions evoking phantom sensations in representative upper and lower limb amputees and mastectomy patients. a, Stimuli to green, lined areas were referred to the other fingers. Stimuli to circles areas were referred to the entire hand. b, stimuli to the blue lined ares were mainly referred to the big toe; stimuli to red circled areas were referred to the entire foot; stimuli to the green lined areas were referred to the other fingers. c, stimuli to circled areas were referred to the phantom nipple.

The issue of the representation of the nipple in the human somatosensory system has not been addressed directly. Thus, although phantom nipple sensations evoked by stimuli on the pinna suggest anatomical connections between the two body parts, no clear-cut interpretation of this result can be provided at present. Given the posited representational contiguity in the somatosensory system of hand and face, of leg and foot, and of foot and genitals (Penfiled and Rasmussen 1950; Woolsey et al. 1979), the evoked phantom sensations in limb amputees have been considered as possible markers of the massive neural rearrangement (Ramachandran et al. 1992; Halligan et al. 1993; Aglioti et al. 1994a, b) reported in non-human primates (Pons et al. 1991; Merzenich and Sameshima 1993; Weinberger 1995; Florence et al. 1997).

Studies addressing directly the issue of remapping in amputees have shown conspicuous somatosensory and motor reorganization (Yang et al. 1994; Kew et al. 1994, 1997; Pascual-Leone et al. 1996) that may indicate the extent to which sensory inputs from the face and fingers have re-occupied deprived somatosensory regions originally subservient the amputated body parts. Magnetoencephalography studies, however, suggest that the amount of remapping may not be related to the type and quality of evoked phantom sensations but rather reflect phantom limb perceptions (Flor et al. 1995; Knecht et al. 1996).

Although, perceptual studies cannot give any direct evidence of the plastic neural changes that may occur as a consequence of a deprivation, the analysis of temporal changes of evoked phantom sensations may provide clues to the possible mechanisms of neural rearrangements. We addressed this issue in a follow-up study of a left-index amputee who experienced vivid, mostly unpleasant, spontaneous phantom sensations (Aglioti et al. 1997). About six months after the amputation, tactile stimulations of either the left hemiface or the third, fourth and fifth fingers of the left, mutilated hand revealed the existence of orderly topographic maps of the missing index on both face and remaining fingers. About three years later, spontaneous and evoked phantom sensations were still present, and the orderly topographic maps of the amputated index on the ipsilateral fingers were unchanged, but the map of the index on the ipsilateral hemiface had totally disappeared. Some phantom sensations could now be aroused by touching the right hemiface, but there was no precise correspondence in kind, intensity and location between the facial stimuli and the sensations evoked by them on the phantom finger (Fig. 3). Similar side shifts and degradation over time of facial maps of a phantom have also been observed by Halligan and co-workers (1994) in an arm amputee.

The different destiny of the orderly maps of the phantom index on the remaining fingers, which persisted for years, and on the face, which did not persist, may depend on very different kinds of associations occurring between veridical and phantom sensations during ordinary behaviour. In accord with typical reports from other amputees, our subject attested that he was particularly aware of sensations from the phantom index when he manipulated objects with the mutilated hand, when phantom and veridical sensations from the remaining fingers were so congruent and correlated that he had the impression that the hand was intact and that his motor commands were directed to the missing finger as much as to the other fingers. By contrast, although he knew that facial stimuli could elicit sensations in the phantom index, he was rarely if at all aware of such sensations in everyday life, in agreement with the apparent irrelevance for efficient behaviour and cognition of an association between veridical facial sensations and phantom sensations in the lost finger.

Thus, time-related changes in the location of skin areas evoking phantom sensations in amputees suggest that the expression of foreign inputs in a deafferented somatosensory cortical area may reflect a hierarchical organisation of different inputs to that area. A somatosensory neuron, for example, may have afferents coming from more than one body district, say the face and the fingers on the

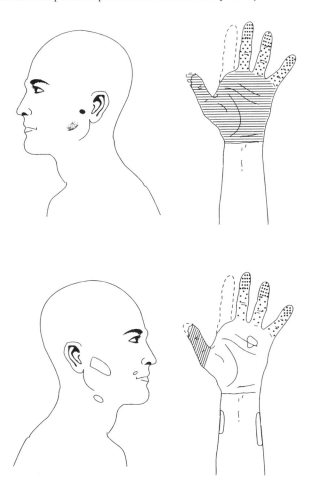

Fig. 3. Follow-up study of the pattern of areas evoking phantom sensation in a left-index amputee patients. Testing sessions carried out six months and three years after the amputation are shown in the upper and lower parts of the figure. Stimuli delivered to dotted areas evoked sensations referred topographically to the phantom index. Stimuli to black and green lined areas were referred to the nail and the last two phalanges of the phantom index, respectively. Stimuli to red areas were referred to the entire phantom index. Circled areas indicate points from where evoked phantom sensations were feeble and non-topographic. Black, lined areas on the hand were not stimulated because the were covered with a bandage.

same side of the body. The more or less strong commitment of a neuron to a given body part (say the index) may simply depend upon the number and synaptic weight of afferent fibers from that body part. Neurons become dedicated to the index because more dense and powerful connections come from this finger. These neurons also have connections from the thumb that, although less dense and powerful than inputs from the index, still have high functional relevance. Afferent fibers to the "index neurons" originating from the face are less numerous and functionally less relevant than those arising from the thumb. Callosal inputs from the opposite hemiface may be even less powerful but still present. Finally, afferents from unrelated body parts, for example the big toe, may not even reach the index neurons. It is plausible that, in the absence of any perturbation, dominant inputs to the index suppress all the others latent inputs that in turn may be unmasked by a deafferentation. This unmasking may take place according to hierarchies that reflect the functional relatedness of the different body parts. Thus, although the notion of a somatotopic mapping of the body may still hold true, a more dynamic view is likely to catch in deeper detail the logic of the functional connectivity in the brain. This view is in keeping with recent evidence for a large, sensorimotor representational overlap of the different body parts (Schieber and Hibbard 1993; Sanes et al. 1995; Godde et al. 1995) and for the existence of latent inputs in the somatosensory cortex (Schroeder et al. 1995). This view is also in keeping with the demonstration that deafferented portions of the primary somatosensory cortex in experimental animals become responsive to sensory inputs normally routed to adjacent regions (Pons et al. 1991).

Furthermore, the notion of dominant and latent inputs (Schroeder et al. 1995; Xing and Gernstein 1996) provides a plausible interpretation of why, in the patient with the amputation of the index finger reported by Aglioti et al. (1997), the maps on the remaining fingers appeared more long-lasting than the facial maps. Removal of the dominant input from the somatosensory representation of the index finger may have unmasked the latent inputs from the ipsilateral fingers and hemiface, which in turn may have kept the callosal inputs mostly silent. The absence of functional implications of ipsilateral facial maps may have induced a further loosening of the connectivity power, leaving open the expression of other connections. Shifts of areas evoking phantom sensations on the other side of the face may reflect the amount of callosal connections that are more dense for the axial and proximal parts of the body than for the extremities (Berlucchi 1990). Were this interpretation correct, the shift of the facial map may have to do with transcallosal remapping effects reported in the monkey somatosensory system (Calford and Tweedale 1990; Clarey et al. 1996). While the functional associations between veridical and phantom digital sensations would favour the stabilisation of the inputs from the remaining fingers, the functional insignificance for behaviour of the input from the ipsilateral hemiface would result in a loss of its synaptic efficacy. The consequent expression of the unmasked callosal input from the opposite hemiface, attested by the appearance of weak phantom sensations evoked from that hemiface, would also be doomed to disappear because of its behavioural irrelevance.

Conclusions

The putative mechanisms of the body schema exhibit stability, inasmuch as some brain regions seem to be permanently committed to represent the corresponding body parts in conscious awareness. Spontaneous and evoked phantom sensations can only arise from central neural activations originally linked to the representation of the lost body part in consciousness. A persistent functional link between regions of the primary somatosensory cortex and bodily regions is proven by the effects of direct electrical stimulation during neurosurgical operations under local anaesthesia. In a patient who had lost his right arm almost a quarter century before the operation and had not experienced phantom phenomena for many years, stimuli applied to the standard hand and arm representations in the left somatosensory cortex brought back into his conscious experience long losts sensations from the missing hand and arm (Ojemann and Silbergeld 1995). Moreover, the phantom limb experiences reported by phocomelic children born without one or more limbs (Melzack et al. 1997) suggest that the brain may be genetically predisposed to represent a prototypical human body, regardless of the correspondence or lack thereof between this ideal model and the actual body.

Nevertheless, there is plasticity in the adult brain, insofar as inputs from intact body parts become able to activate deafferented cortical regions. Animal studies indicate that skill learning with a body part leads to an increased representation of that body part in the somatosensory (Recanzone et al. 1992) and motor cortex (Nudo et al. 1997), and that the total loss of a sense modality increases the cortical representation and the functional performance of other sensory modalities (Rauschecker 1995). A certain degree of plasticity is also suggested by the extension of body representation to external objects that acquire, over time, perceptual relation with the body itself (Aglioti et al. 1996).

Acknowledgment

This chapter was prepared with the financial support of M.U.R.S.T. and the Consiglio Nazionale delle Ricerche, Italy.

References

Aglioti S, Bonazzi A, Cortese F, Cugola L, Testoni F (1994a) La percezione dell'arto fantasma come indicatore di plasticita' nel cervello umano adulto. Eur Med Phys 30: 21–28

Aglioti S, Bonazzi A, Cortese F (1994b) Phantom lower limb as a perceptual marker of neural plasticity in the mature human brain. Proc R Soc Lond B 255: 273–278

Aglioti S, Cortese F, Franchini C (1994c) Rapid sensory remapping in the adult human brain as inferred from phantom breast perception. NeuroReport 5: 473–476

Aglioti S, Smania N, Manfredi M, Berlucchi G (1996) Disownership of left hand and objects related to it in a right-brain-damaged patient. NeuroReport 8: 293–296

Aglioti S, Smania N, Atzei A, Berlucchi G (1997) Spatio-temporal properties of the pattern of evoked phantom sensations in a left index amputee patient. Behav Neurosci 111 (5):

Appenzeller O, Bicknell JM (1969) Effects of nervous system lesions on phantom experience in amputees. Neurology 19: 141–146

Berlucchi G (1990) Commissurotomy studies in animals. In: Boller F, Grafman J (eds) Handbook of neuropsychology. Vol. 4. Amsterdam, Elsevier, 9–47

Bisiach E, Geminiani G (1991) Anosognosia related to hemiplegia and hemianopia. In: Prigatano GP, Schacter DL (eds) Awareness of deficit after brain injury. Oxford University Press, Oxford, 17–39

Bisiach E, Rusconi ML, Vallar G (1991) Remission of somatoparaphrenic delusion through vestibular stimulation. Neuropsychologia 29: 1029–1031

Bonda E, Petrides M, Frey S, Evans A (1995) Neural correlates of mental transformations of the body-in-space. Proc Natl Acad Sci USA 92: 11180–11184

Calford MD, Tweedale R (1990) Interhemispheric transfer of plasticity in the cerebral cortex. Science 249: 805–807

Clarey JC, Tweedale R, Calford MD (1996) Interhemispheric modulation of somatosensory fields: evidence for plasticity in primary somatosensory cortex. Cereb Cortex 6: 196–206

Critchley M (1979) The divine banquet of the brain. New York, Raven Press, 92–105

Cronholm B (1951) Phantom limbs in amputees. Acta Psych Neurol Scand suppl 2, 1–310

Flor H, Elbert T, Knecht S, Weinbruck C, Pantev C, Birbaumer N, Larbig W, Taub E (1995) Phantom-limb pain as a perceptual correlate of cortical reorganization following arm amputation. Nature 375: 482–484

Florence SL, Jain N, Kaas JH (1997) Plasticity of somatosensory cortex in primates. Seminars Neurosci 9: 3–12

Frederiks JAM (1985a) Body schema disorders. In: Frederiks JAM (ed) Handbook for clinical neurology. Vol. 45 (1). Amsterdam, Elsevier, 373–393

Frederiks JAM (1985b) Phantom limbs and phantom pain. In: Frederiks JAM (ed) Handbook of clinical neurology. Vol. 45 (1). Amsterdam, Elsevier, 395–404

Godde B, Hilger T, von Seelen W, Berkefeld T, Dinse HR (1995) Optical imaging of rat somatosensory cortex reveals representational overlap as topographic principle. NeuroReport 7: 24–28

Halligan PW, Marshall JC, Wade DT, Davey J, Morrison D (1993) Thumb in cheek? Sensory reorganization and perceptual plasticity after limb amputation. NeuroReport, 4: 233–236

Halligan PW, Marshall JC, Wade DT (1994) Sensory disorganization and perceptual plasticity after limb amputation: a follow-up study. NeuroReport 5: 1341–1345

Halligan PW, Marshall JC, Wade TD (1995) Unilateral somatoparaphrenia after right hemispheric stroke: a case description. Cortex 31: 171–182

Head H, Holmes G (1911) Sensory disturbances from cerebral lesions. Brain 34: 102–254

Iriki A, Tanaka M, Iwamura Y (1996) Coding of modified body schema during tool use by macaque postcentral neurones. NeuroReport 7: 2325–2330

Kaas JH (1991) Plasticity of sensory and motor maps in adult mammals. Ann Rev Neurosci 14: 137–167

Katz J (1992) Psychophysiological contributions to phantom limbs. Can J Psychiat 37: 282–298

Katz J, Melzack R (1990) Pain "memories" in phantom limbs: review and clinical observations. Pain 43: 319–336

Kew JJM, Ridding MC, Rothwell JC, Leigh PN, Passingham RE, Sooriakumaran S, Frackowiak RSJ, Brooks DJ (1994) Reorganization of cortical blood flow and transcranial magnetic stimulation maps in human subjects after upper limb amputation. J Neurophysiol 72: 2517–2524

Kew JJM, Halligan PW, Marshall JC, Passingham RE, Rothwell JC, Ridding MC, Marsden CD, Brooks DJ (1997) Abnormal access of axial vibrotactile input to deafferented somatosensory cortex in human upper limb amputees. J Neurophysiol 77: 2753–2764.

Knecht S, Henningsen H, Elbert T, Flor H, Höhling C, Pantev C, Taub E (1996) Reorganizational and perceptual changes after amputation. Brain 119: 1213–1219

Lackner JR (1988) Some proprioceptive influences on the perceptual representation of body shape and orientation. Brain 111: 281–297

Levine DN, Calvanio R, Rinn WE (1991) The pathogenesis of anosognosia for hemiplegia. Neurology 41: 1770–1781

McGilchrist I, Cutting J (1995) Somatic delusions in schizophrenia and the affective psychoses. Br J Psychiatry 167: 350–361

McGlynn SM, Schacter DL (1989) Unawareness of deficits in neuropsychological syndromes. J Clin Exp Neuropsychol 11: 143–205

Melzack R (1990) Phantom limbs and the concept of neuromatrix. Trends in Neurosci 13: 88–92

Melzack R (1992) Phantom limbs. Sci Am 266 (4): 90–96

Melzack R, Israel R, Lacroix R, Schultz G (1997) Phantom limbs in people with congenital limb deficiency or amputation in early childhood. Brain 120: 1603–1620

Merzenich MM, Sameshima K (1993) Cortical plasticity and memory. Curr Opin Neurobiol 3: 187–196

Morselli E (1886) Sulla dismorfofobia e sulla tafe fobia. Bollettino della regia Accademia di Genova VI: 110–119

Moss AD, Turnbull OH (1996) Hatred of the hemiparetic limbs (misoplegia) in a 10 year old child. J Neurol Neurosurg Psychiat 61: 210–211

Nudo RJ, Platz EJ, Milliken GW (1997) Adaptive plasticity in primate motor cortex as a consequence of behavioral experience and neuronal injury. Seminars Neurosci 9: 13–23

Ojemann GJ, Silbergeld DL (1995) Cortical stimulation mapping of phantom limb rolandic cortex. J Neurosurg 82: 641–644

Pascual-Leone A, Peris M, Tormos JM, Pascual-Leone Pascual A, Catalá MD (1996) Reorganization of human cortical motor output maps following traumatic forearm amputation. NeuroReport 7: 2068–2070

Penfield W, Jasper H (1954) Epilepsy and the functional anatomy of the human brain. Boston, Little, Brown & Co.

Penfield W, Rasmussen T (1950) The cerebral cortex of man: a clinical study of localization of function. New York, McMillan Press

Pons T, Garraghty PE, Ommaya AK, Kaas JH, Taub E, Mishkin M (1991) Massive cortical reorganization after sensory deafferentation in adult macaques. Science 252: 1857–1860

Ramachandran VS (1995) Anosognosia in parietal lobe syndrome. Consciousness Cognition 4: 22–51

Ramachandran VS, Stewart M, Rogers-Ramachandran DC (1992) Perceptual correlate of massive cortical reorganization. NeuroReport 3: 583–586

Rauschecker JP (1995) Compensatory plasticity and sensory substitution in the cerebral cortex. Trends Neurosci 18: 36–43

Recanzone GH, Merzenich MM, Jenkins WM (1992) Frequency discrimination training engaging a restricted skin surface results in an emergence of a cutaneous response zone in cortical area 3a. J Neurophysiol 67: 1057–1070

Rode G, Charles N, Perenin MT, Vighetto A, Trillet N, Aimard G (1992) Partial remission of hemiplegia and somatoparaphrenia through vestibular stimulation in a case of unilateral neglect. Cortex 38: 203–208

Roper SN, Levesque MF, Sutherling WW, Engel J Jr (1993) Surgical treatment of partial epilepsy arising from the insular cortex. Report of two cases. J Neurosurg 79: 266–269

Sanes JN, Donoghue JP, Thangarai V, Edelman RR, Warach S (1995) Shared neural substrates controlling hand movements in human motor cortex. Science 268: 1775–1777

Schieber MH, Hibbard LS (1993) How somatotopic is the motor cortex hand area? Science 261: 489–492

Schroeder CE, Seto S, Arezzo JC, Garraghty PE (1995) Electrophysiological evidence for overlapping dominant and latent inputs to somatosensory cortex in squirrel monkeys. J Neurophysiol 74: 722–732

Weinberger NM (1995) Dynamic regulation of receptive fields and maps in the adult sensory cortex. Ann Rev Neurosci 18: 129–158

Woolsey CN, Erickson TC, Gilson W (1979) Localization in somatic sensory and motor areas of human cerebral cortex as determined by direct recording of evoked potentials and electrical stimulation. J Neurosurg 51: 476–506

Xing J, Gerstein GL (1996) Networks with lateral connectivity. III. Plasticity and reorganization of somatosensory cortex. J Neurphysiol 75: 217–232

Yang TT, Gallen CC, Ramachandran VS, Cobb S, Schwartz BJ, Bloom FE (1994) Non-invasive detection of cerebral plasticity in the adult human somatosensory cortex Neuroreport, 5, 701–704

Young AW, Leafhead KM, Szulecka TK (1994) The Capgras and Cotard delusions. Psychopathology 27: 226–231

Neuroplasticity in the Adjustment to Blindness

A. Pascual-Leone[1,2], R. Hamilton[1], J. M. Tormos[2], J. P. Keenan[1]
and *M.D. Catalá[2]*

Summary

Loss of vision due to injury to the eyes results in deafferentation of very large areas of the human cortex and poses striking demands on other sensory systems to adjust to blindness in a society that heavily relies on vision. Blind subjects need to extract crucial spatial information from touch and hearing. To accomplish this, plastic trans-modal changes appear to take place by which a larger area of the sensorimotor cortex is devoted to the representation of the reading finger in Braille readers, and parts of the former visual cortex are recruited for the processing of tactile and auditory information.

These findings provide evidence of trans-modal sensory plasticity in humans. Similar mechanisms might be involved in other forms of skill learning and recovery from lesions. Recent studies suggest the possibility that available neurophysiologic techniques might not only be used to reveal such plastic changes, but may also have a potential role in guiding the plastic changes, thus improving functional outcome.

Introduction

Braille reading provides blind subjects with the opportunity to read and communicate in writing, thus greatly expanding their integration into society and their opportunities for employment. Braille reading requires the processing of tactile information into meaningful shapes. Subjects have to discriminate small patterns of raised dots with the pad of their index finger and extract spatial information through touch rather than vision. To accomplish this task, subjects move their index finger from side to side at a controlled speed so that the sensory skin receptors are maximally activated by the raised dots and as much information as possible is obtained. Therefore, learning to read Braille poses a great demand on sensory and motor representations from a rather small part of the body, the dis-

---

[1] Laboratory for Magnetic Brain Stimulation, Dept. Neurology, Beth Israel Deaconess Medical Center & Harvard Medical School, Boston MA, USA
[2] Unidad de Neurobiologia, Dept. Fisiologia, Universidad de Valencia and Instituto Cajal, Consejo Superior de Investigaciones Científicas, Spain

J. Grafman / Y. Christen (Eds.)
Neuronal Plasticity:
Building a Bridge from the Laboratory to the Clinic
© Springer-Verlag Berlin Heidelberg New York 1999

tant pad of the index finger, and stresses the spatial coding capabilities of tactile exploration.

Blind Braille readers do not have a lower peripheral sensory threshold in the pad of the reading index finger than blind, non-Braille reading controls or sighted volunteers (Pascual-Leone and Torres 1993). However, the sensory and motor representations of the reading index finger in the brain are significantly larger for Braille readers than for blind or sighted non-Braille readers (Pascual-Leone and Torres 1993; Pascual-Leone et al. 1993). This enlarged sensorimotor representation in blind Braille readers develops slowly, over the course of months (Pascual-Leone et al. 1993, 1996b) and is modulated by the preceding activity (Pascual-Leone et al. 1995c). Therefore, changes in representation of the reading finger in the brain cortex seem to play a critical role in the acquisition of the Braille reading skill.

In addition, both acquired and congenitally blind subjects demonstrate activation in their primary and secondary visual cortices when reading Braille (Sadato et al. 1996). Participation of the striate "visual" cortex in a tactile task seems related to the difficulty of the tactile discrimination, regardless of whether there is a lexical component to it or not (Sadato et al. 1996). In the congenitally blind, interference with the function of the occipital cortex during Braille reading using repetitive transcranial magnetic stimulation (rTMS) results in disruption of the Braille reading skill (Cohen et al. 1997). Therefore, it appears that Braille reading in the blind is an example of true "trans-modal sensory plasticity," by which the deafferented, formally visual cortex is recruited for demanding tactile tasks, thus enhancing the sensory discrimination abilities of the blind subject and making the acquisition of the tactile Braille reading skill possible.

In summary (Fig. 1), early blind subjects who were born blind or became blind before age seven seem to develop, as they learn to read Braille, a larger sensorimotor representation of the reading finger and a recruitment of the occipital cortex for the tactile discrimination task. Therefore, the neural network involved in spatial decoding of tactile information appears different in early blind subjects than in late blind subjects and sighted controls.

These findings lend support to the notion that, faced with changing demands following injury or during the acquisition of new skills, striking plastic changes can take place in the human nervous system. Cortical areas can be expanded and new elements can be recruited to functional neural networks. Training and functional demand are likely to play crucial roles in the facilitation of these plastic processes. In the case of blind subjects, motivation is enormously high since the expected benefits for social integration are considerable. Will, desire, perseverance, and expected functional gain are difficult to quantify, but likely to be equally important factors.

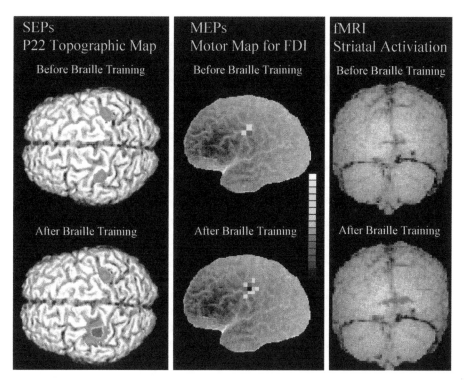

Fig. 1. Evidence of plasticity in sensory, motor and occipital cortices in early blind subjects after learning Braille. Representative examples of the studies performed before and at the end of one year of learning Braille are shown. The different studies were conducted on different subjects using somatosensory evoked potentials (SEPs) to mechanical stimuli to the index finger pad; motor mapping with transcranial magnetic stimulation of the potentials evoked in the first dorsal interosseus muscle (FDI, side-to-side mover of the index finger), and BOLD functional magnetic resonance imaging (fMRI) while reading Braille characters

Enlarged Sensorimotor Representation of the Reading Finger in Blind Braille Readers

A growing body of evidence shows the dynamic, plastic properties of the adult central nervous system in response to injury, activity and skill acquisition (Merzenich et al. 1990; Merzenich and Jenkins 1993; Kaas 1991). Specifically, a large number of studies support the view that somatosensory representation can be selectively remodeled by new tactile experiences in intact adult individuals. In monkeys, when tactile stimuli are applied repetitively to restricted skin regions, representational maps of the stimulated body part and its cutaneous receptive field size have been shown to increase substantially (Jenkins et al. 1990). This remodeling has been shown to relate to an improvement in tactile discrimination abilities and to be heavily influenced by the potential benefit derived from the new skill (Recanzone et al. 1992 a–d)

In humans, motor and sensory evoked potential studies similarly show that the sensorimotor representation of the reading finger is enlarged in blind, proficient Braille readers compared with the sensorimotor representation of the homologous finger of their other hand or with either finger of sighted and blind, non-Braille reading controls (Pascual-Leone and Torres 1993; Pascual-Leone et al. 1993).

In blind, proficient readers, Braille reading shows consistent activation in the primary sensorimotor cortex (SM1) bilaterally on functional neuroimaging studies (Sadato et al. 1996). Bilaterality of the activation might be related to the complexity of the Braille reading task. However, activation of SM1 contralateral to the reading finger is more prominent than ipsilateral activation, particularly for right-handed reading subjects (Sadato et al. 1996). These findings are consistent with results from animal studies, revealing a more complex and extended cortical sensorimotor representation of the preferred hand and a correlation between activity and size of the cortical representation (Nudo et al. 1992). In learning tactile Braille, the SM1 contralateral to the preferred finger might show greater activation due to increased excitability related to increased use. This notion is consistent with the rapid modulation of the cortical output maps in proficient blind Braille readers depending on preceding activity (Pascual-Leone et al. 1995c) and other studies of modulation of motor cortical outputs during skill acquisition (Pascual-Leone et al. 1994b, 1995b).

What is the Time Course of this Enlargement of the Sensorimotor Representation of the Reading Finger when Subjects are Learning Braille?

We have been able to study 10 blind subjects serially over the course of one year as they learned to read Braille. The subjects (ages 13 to 29 years, mean age 18.5 years) had become blind one to six years prior to the study, due to traumatic injury to the eyes (hunting or traffic accidents). None had a progressive neurological or systemic disease. All had normal neurological exams aside from the blindness and had normal cerebral MRI studies. The subjects were integrated in courses imparted by the Spanish Organization for the Blind (ONCE) to teach Braille reading skills to blind subjects. These courses followed a variable period of rehabilitation during which the subjects had already become familiar with the Braille symbols and had acquired basic Braille reading skills. Therefore, these courses were designed to help the students improve their Braille reading skills to aide them in their academic ambitions. Classes (two hrs/day) were held five days a week (Monday to Friday). The course lasted from September to June and was interrupted in December for two weeks around Christmas and in March/April for two weeks around Easter. During the weekends, all subjects were encouraged by their teachers to practice their Braille reading skill, but they all reported spending very little time reading. However, during the week, in addition to the class, they all practiced one to two hours daily to complete their "homework."

We used TMS to map the cortical motor outputs to the first dorsal interosseus muscle (FDI), the prime agonist for the side-to-side movement required during proficient Braille reading. The mapping technique is described in detail elsewhere (Pascual-Leone and Torres 1993; Pascual-Leone et al. 1993, 1995c). The changes in cortical output maps to the FDI seem to provide reliable information about changes in sensorimotor representation of the index finger. Maps were obtained on Mondays, before the weekly classes and after the weekend rest, and on Fridays at the end of the class, thus at the end of the week of daily classes. This design was chosen based on our previous findings regarding the variability of the cortical output maps depending on preceding activity (Pascual-Leone et al. 1995c).

Figure 2 summarizes the findings in all subjects. Over the course of 10 months the subjects' Braille reading skill improved significantly, as demonstrated by the increasing number of words read per minute. The cortical output maps to the non-reading hand did not change over this period. However, striking changes

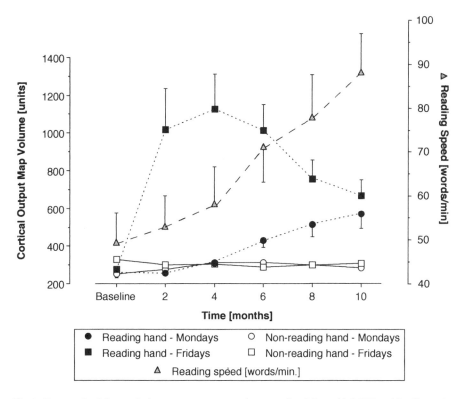

Fig. 2. Line graph of the cortical motor output map volumes to the right and left FDI and Braille reading speed over the course of 10 months of Braille learning. Results display mean ± standard deviation values for 10 subjects. Monday and Friday maps refer to the time of testing in relation to the week of daily Braille classes

were noted in the volume of the cortical motor output map to the FDI of the reading hand. These changes seemed to follow different time courses for the Fridays (after the week of classes) and Mondays (after the weekend rest and prior to the classes). Friday maps changed rapidly and dramatically from the beginning of the study, continued to increase in volume for six months and then decreased back towards baseline. On the other hand, Monday maps did not change until the sixth month of learning, then increased slowly and plateaud by the tenth month. The behavioral data regarding Braille reading speed seemed correlated more tightly with the Monday than with the Friday map changes.

The different temporal profiles for changes in cortical output maps for the Friday and Monday maps suggest different underlying mechanisms. On Fridays the changes were demonstrated early and were dramatic; however, they appeared to go back to baseline by the following Monday after two days of rest. On the other hand, the Monday maps changed much more slowly, but when they eventually did, the enlargement was less prominent but more stable. At the end of the 10 months of classes subjects went on vacation, but upon their return three months later, the cortical motor output maps to the FDI were unchanged from the last Monday map and significantly larger than at the beginning of the study.

We would argue that the changes demonstrated reflect fast, transient modulation of existing connections and eventual new structural changes. The changes demonstrated by the Friday maps are likely due to activity-modulated unmasking of connections or increases in synaptic efficacy. On the other hand, the changes in Monday maps took long enough and were stable enough to support the possibility of sprouting and structural changes. Therefore, this study suggests that, in the acquisition of the Braille reading skill, as presumably in the acquisition of other skills, there are transient, rapid changes in efficacy of existing connections that lead the way for eventual structural changes in face of continued practice.

Role of the Occipital Cortex in Braille Reading in the Early Blind

Proficient Braille reading by blind subjects activates the dorsal and ventral portion of the occipital cortex (Sadato et al. 1996). Furthermore, there is suppression of parietal operculum and activation of the ventral portion of the occipital cortex in blind subjects during tactile discrimination tasks, contrary to the pattern of activation observed in the sighted subjects (Sadato et al. 1996). Studies by Uhl et al. (1991, 1993) using event related potentials and cerebral blood flow measures also suggest occipital cortex activation in early blind humans. These findings suggest that the pathway for tactile discrimination changes with blindness.

In a recent follow-up study, we found that repetitive TMS applied to the occipital cortex was able to disrupt processing of tactile spatial information – reading of embossed Roman characters – only in early blind subjects (subjects born blind or that had become blind before the age of seven; Cohen et al. 1997). In this study, subjects were aware that they had felt Braille characters, but during

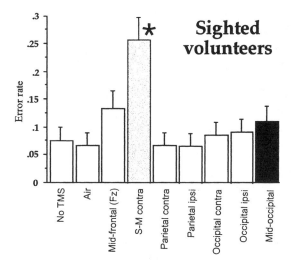

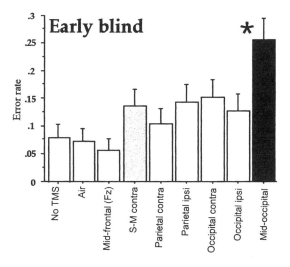

Fig. 3. Number of errors in discrimination of Braille characters during exposure to repetitive TMS in early blind and sighted subjects depending on the site of rTMS application. The asterisks mark significant differences between the early blind subjects and the sighted controls. Adapted from Cohen et al. (1997)

rTMS to their occipital cortex they were unable to discriminate them (Fig. 3). This effect was limited to early blind subjects. In sighted subjects, rTMS to the visual cortex does not interfere with the ability to detect or discriminate embossed Roman letters by touch. In a subsequent study, Cohen et al. (unpublished data, personal communication) have found that, in late blind subjects, rTMS to the striate cortex does not interfere with reading of embossed Roman characters either. In the early blind subjects, rTMS to the occipital cortex does not result in failure of detection, but rather interferes with perception of the Braille stimuli. During occipital rTMS, early blind subjects know that they are touching Braille symbols, but the Braille dots "feel different," "flatter," "less sharp

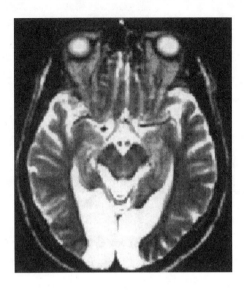

Fig. 4. T2-weighted MRI of the bilateral occipital stroke in an early blind patient resulting in Braille alexia

and less well defined." Occasionally, some subjects even report feeling additional ("phantom") dots in the Braille cell, and they are unable to correctly identify the Braille symbol or Roman embossed letter presented.

These findings support the hypothesis that detection of tactile stimuli does not require striate cortex, but perception (i.e., discrimination of the tactile stimuli) does involve striate information processing, though only in the early blind. Further support for the role of striate cortex in the perception of complex tactile stimuli comes from the findings in a patient whom we recently studied. She had been blind since birth due to premature retinopathy. She was a highly proficient Braille reader who used Braille at her work. She was a proof reader for the newsletter for the Spanish Organization for the Blind for four to six hours per day. At age 62 she had an acute neurological event that resulted in coma from which she recovered, but she was left unable to read Braille. She had no difficulties discriminating textures or identifying everyday objects by touch, but she was greatly impaired in tactile discriminations that required complex spatial decoding. She made frequent errors trying to identify even single Braille letters and she was completely unable to read Braille words. She described being able to "feel" the Braille dots, but she could not "make sense" of what she was touching. She complained that it felt "as if I had never learned Braille at all." Her neurological exam was otherwise unremarkable, and her MRI showed a bilateral occipital stroke (Fig. 4). Therefore, the findings in this case, emphasize the role that the occipital cortex plays in Braille reading in early blind subjects.

What is the Flow of Spatial Information Processing from Somatosensory Input to Occipital Cortex in the Blind?

TMS can provide information about the timing of information processing along a neural network (Maccabee et al. 1991), evaluate the functional significance of elements of a neural network in a given task, thus enhancing the information derived from neuroimaging studies (Cohen et al. 1997), and be combined with neuroimaging studies to demonstrate the functional connectivity between cortical areas (Paus et al. 1997). TMS can transiently disrupt the function of a targeted cortical region and thus, applied at variable intervals following a given stimulus, it can provide information about the temporal profile of activation and information processing along elements of a neural network (Pascual-Leone et al 1994 c, 1997 a). This application has been used in normal volunteers to study the timing of visual information processing in the striate cortex (Maccabee et al. 1991) and of somatosensory information in the primary sensory cortex (Cohen et al. 1991).

A transcranial magnetic stimulus appropriately delivered in time and space can transiently disrupt the arrival of the thalamo-cortical volley into the primary sensory cortex and interfere with detection of peripheral somatosensory stimuli (Cohen et al. 1991). This disruptive effect will result in the subject's failure to detect the stimulus. The subject is not aware that he or she received a peripheral somatosensory stimulus prior to TMS. To achieve this effect, the TMS cortical stimulus (CS) has to be appropriately timed following the peripheral stimulus (PS) and the current induced in the brain appropriately oriented (Pascual-Leone et al. 1994 a). Detection of PS is disrupted only when the interval between PS and CS is 15 to 35 ms (Cohen et al. 1991; Pascual-Leone et al. 1994 a). In addition, topographic specificity can be demonstrated according to the known somatotopic organization of the sensory cortex. TMS must be delivered at the appropriate site for projection of index finger afferences when PS is applied to the index finger pad, and no effect is demonstrable if the site of TMS is displaced by 1 or 2 cm in any direction (Pascual-Leone et al. 1994 a).

These findings provide information about the time course of information arrival to the primary sensory cortex and its processing time there in normal subjects. The same effect of blocking detection of somatosensory stimuli can be demonstrated in blind proficient Braille readers (Pascual-Leone and Torres 1993). In 1993 we reported that detection of electric stimuli applied to the pad of the index finger could be blocked by properly timed transcranial magnetic stimuli to the contralateral sensorimotor cortex. Using a specially designed stimulator that resembled a Braille cell, we applied electric stimuli slightly above sensory threshold to the index finger pad of the right or left hand in sighted controls and blind subjects. These peripheral stimuli were followed by TMS stimuli at variable intervals and intensities to different scalp positions targeting the sensorimotor cortex. TMS stimuli appropriately delivered in time and space resulted in a block of detection of the peripheral stimuli, such that the subjects were unaware of having received a peripheral stimulus preceding the cortical stimulus. In blind subjects, the block of detection of the peripheral stimulus could be achieved by cortical stimuli delivered 1) to a larger cortical area and 2) at a longer window of intervals between peripheral and cortical stimuli, but required a stronger intensity of the TMS stimuli.

We have recently used a similar approach to evaluate the timing and contributions of somatosensory *and occipital cortex* to processing of tactile information in the blind. We have studied three subjects who were blind due to premature retinopathy and had no residual vision with absent responses to visual evoked potentials. All three were right-handed males, ages 43, 49, and 51 years. All three had normal neurological exams, except for the blindness, and normal brain MRIs. All three were proficient Braille readers and used Braille one to six hours per day.

Real or non-sensical Braille stimuli were presented with a specially designed Braille stimulator to the pad of the reading index finger. All three subjects used primarily their right index finger for Braille character recognition. Single-pulse TMS stimuli were applied to the left or right sensorimotor cortex and the striate occipital cortex at variable intervals following the presentation of the Braille stimuli. The site of stimulation was determined by correlating the scalp position of the TMS coil with the subject's 3D reconstructed MRI using a frameless stereotactic targeting device. The study was conducted in blocks of 360 trials. Each subject completed three blocks of 360 trials that differed in the cortical area targeted by TMS. In each block three Braille stimulation conditions were tested (no Braille stimulus, real Braille stimulus, and non-sensical Braille stimulus). In addition, in each block 12 different TMS conditions were tested (no TMS and 10–110 ms interstimulus intervals between Braille stimulus and TMS).

Figure 5 summarizes the experimental design schematically (A) and the results for all three subjects graphically (B).

This study allows the following conclusions: 1) TMS to the left somatosensory cortex disrupted detection of real and non-sensical Braille stimuli at interstimulus intervals of 20 to 40 ms. The subjects did not realize that a peripheral stimulus had been presented. In the instances in which they did realize the presentation of a peripheral stimulus, they were able to correctly identify whether it was real Braille or not and what Braille symbol was presented.

2) TMS to the striate cortex disrupted the processing of the peripheral stimuli at interstimulus intervals of 50 to 80 ms. Contrary to the findings after sensorimotor TMS, the subjects generally knew whether a peripheral stimulus had been presented or not, therefore no interference with detection was demonstrated. However, the subjects were unable to discriminate whether the presented stimuli were real or non-sensical Braille or what Braille symbol might have been presented (interference with perception).

Therefore, in early blind subjects, the interval between a tactile stimulus to the finger pad and a cortical stimulus that interferes with processing of tactile information is different for cortical stimulation of the somatosensory and the striate ("visual") cortex. This time difference provides insight into the temporal profile of information processing and transfer in early blind subjects between somatosensory and striate cortex. Theoretically, two main alternative routes

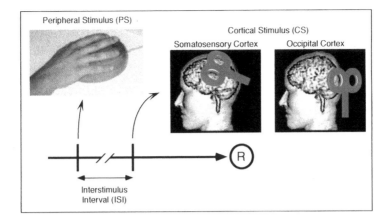

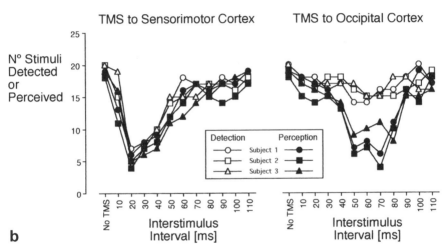

Fig. 5. Summarizes the experimental design schematically (**A**) and the results for all three subjects graphically (**B**). The graph displays in open symbols the number of stimuli detected by each subject depending on TMS condition and regardless of whether real or non-sensical Braille stimuli were presented. The graph displays in filled symbols the number of correctly identified Braille stimuli (real versus non-sensical and what Braille character) by each subject depending on TMS condition from among the stimuli that had been detected in the first place

could be entertained: 1) thalamo-cortical connection to sensory and visual cortex and 2) cortico-cortical connections from sensory cortex to visual cortex.

Thalamic somatosensory nuclei could send input to both the somatosensory cortex and the striatal cortex from multimodal cells in the geniculate nuclei. These theoretical multiple projections might be masked or even degenerate in the postnatal period, given normal vision. However, in early blind subjects, these somatosensory thalamo-striate projections might remain and be responsible for the participation of the striate cortex in tactile information processing. Murata et al. (1965) demonstrated the existence of weak non-visual input to cells of the newborn cat primary visual cortex. Such tenuous multimodal pathways might be unmasked in the case of early injury and visual deprivation. Unfortunately, no further studies have evaluated the fate of such multimodal input into the visual cortex in adult or blind animals or humans.

Further support for a subcortical origin of the recruitment of the visual cortex for tactile information processing in the early blind can be seen in the changes induced by blindness in the neuronal populations in the geniculate nucleus. Rauschecker (1995) found that early visual deprivation results in an increase in the number of multimodal neurons in the geniculate nucleus. Such multimodal neurons receive and process auditory and tactile information and are presumed to retain their projection to visual cortex.

Nevertheless, cortico-cortical connections between sensory cortex and visual cortex seem the more likely route of recruitment of the striate cortex for tactile spatial information processing in the early blind. A sequential striate-prestriate-inferior temporal cortical pathway (ventral visual pathway) is known to serve visual discrimination functions (Ungerleider and Mishkin 1982). In macaque monkeys, ablation of the posterior part of the inferior temporal cortex (area TEO) leads to severe visual pattern discrimination deficits (Blake et al. 1977). Recent PET studies in humans show activation in the fusiform gyrus in occipital and occipito-temporal cortex during tasks that require attention to form, such as processing of pictures, faces, letter strings and geometric shapes (Corbetta et al. 1990; Haxby et al. 1991). These reports support the notion that the ventral visual pathways is used for visual shape discrimination. Murray and Mishkin (1984) have suggested an analogy between pathways of visual and tactile shape discrimination. The secondary sensory cortex (SII) for touch discrimination may be analogous to the posterior region of the inferior temporal cortex (area TEO) for visual pattern discrimination, and the insula may be analogous to the anterior part of the inferior temporal cortex (area TE). In the monkey, the posterior parietal association cortex (area 7) is interconnected with the visual association cortex (dorsolateral area 19; Bruce et al. 1981). Early visual deprivation in the monkey makes most neurons in area 7 and 19 responsive to somatic exploration (Hyvarinen et al. 1981), and diffuse reciprocal projections link area 19 to the primary visual cortex (Shipp and Zeki 1989). These findings suggest that somatosensory input could be transferred to the primary visual cortex through the dorsal visual association areas during spatial tactile information processing by blind subjects. The spatial information originally conveyed by the tactile modality in the sighted subjects (SI-SII-insular cortex-limbic system) might be processed in the blind by the neuronal networks usually reserved for the visual, shape discrimination process (SI-BA7-dorsolateral BA 19-V1-occipito-temporal region – anterior temporal region – limbic system). This plasticity through cortico-cortical connections would explain the fact that tactile information processing in the somatosensory and occipital cortex in the early blind is not only different in timing but also in the type of contribution – detection versus discrimination or perception, respectively.

Can Braille Learning be Accelerated?

Sighted subjects, even instructors of the blind, who learn to read Braille generally do so using sight rather than touch. Even visually impaired subjects, for example those with cataracts, diabetic retinopathy, or glaucoma, generally do not learn tactile Braille reading until visual loss is very severe. At the Carroll and the Perkins Schools for the Blind in the Boston area, as well as in the schools of the Spanish Association for the Blind (ONCE), partially blind children are sometimes blindfolded during tactile Braille reading classes to speed-up the learning process (personal communication, Medical Director of the Carroll School and Director of the ONCE in Valencia). However, no controlled studies are available to support this practice, which is based largely on "experience" and anecdotal evidence.

For many years in Spain, instructors for the blind underwent intensive teaching at a boarding school from the ONCE in Sabadell. As part of their instruction, students wore a blindfold for one week "in order to experience blindness first hand" (personal communication, Director of the ONCE in Valencia). Subjects who underwent this training related enhanced abilities to derive spatial information from touch over the course of the blindfolded week (Pascual-Leone, interviews of many of the instructors that underwent this training). They all tolerated the procedure without complication. They all noticed improved abilities to orient to sounds and judge distance by sound by the end of the week. Similarly, they all noted improved abilities to differentiate surfaces and identify objects by touch during the seven blindfolded days. Interestingly, at the end of the week, removal of the blindfold required a period of several hours for re-adjustment to a visual world, with initial difficulties in visual spatial information decoding.

These data suggest the possibility that recruitment of the occipital cortex for processing of tactile information is not only functionally relevant in Braille reading, but necessary for the acquisition of the Braille reading skill. If so, can the Braille reading ability in blind subjects be enhanced by strategies that result in an increased excitability of the occipital cortex?

Repetitive TMS at appropriate stimulation intensity and frequency can result in an increase in cortical excitability beyond the duration of the TMS application itself (Pascual-Leone et al. 1994d, 1997b; Tergau et al. 1997). This procedure can be applied to a variety of neuropsychiatric conditions and lead to a lasting modulation of symptoms (Pascual-Leone et al. 1995a, 1996a). In preliminary studies we have applied trains of rTMS to the occipital cortex of five early blind subjects and evaluated the effects of the stimulation on their Braille reading speed (Fig. 6). Following rTMS at parameters that enhance cortical excitability (Tergau et al. 1997), subjects were able to read faster, whereas their reading speed was decreased by rTMS at 1 Hz which has been shown to result in a lasting depression of cortical excitability (Chen et al. 1997). Sham rTMS did not change the Braille reading speed in any of the subjects.

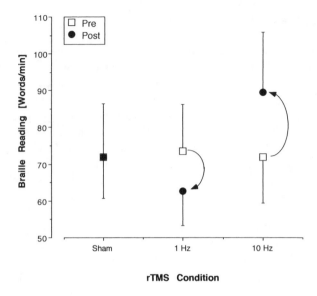

Fig. 6. Line graph of the effects of rTMS (sham, 1 Hz or 10 Hz stimulation) on Braille reading speed measured in words read per min. Pre refers to performance prior to rTMS and Post refers to performance inmediately following rTMS to the occipital cortex. Results display the mean and standard deviation values for five early blind subjects

Conclusions

Blind subjects show a reorganization of the cortical representation of their reading finger when they learn to read Braille. This plastic reorganization appears to involve two distinct processes. First there is a rapid and very striking enlargement of the cortical representation of the reading finger that is likely due to unmasking of connections or upregulation of synaptic efficacy. Later, after six months of Braille learning, there is a slower and less prominent enlargement of the cortical representation of the reading finger that results in more stable changes and might represent structural plasticity. In addition, at least in early blind subjects, the occipital, "visual" cortex appears to play a critical role in Braille reading. Cortico-cortical connections between sensorimotor and occipital cortex might be responsible for the recruitment of the "visual" cortex in tactile information processing. These findings in the case of blind Braille readers ought to be applicable to other situations demanding skill learning and adaptation to neural injury. Finally, rTMS and other strategies to externally modulate cortical excitability might prove useful in enhancing skill acquisition and recovery of function after injury by guiding processes of neural plasticity.

Acknowledgements

Supported in part by grants from the Spanish Ministerio de Educación y Ciencia (DGICYT), the Harvard Thorndike Clinical Research Center (NIH), the Milton Foundation and the National Eye Institute. # ROIEY12091-01

References

Blake L, Jarvis CD, Mishkin M (1977) Pattern discrimination thresholds after partial inferior temporal or lateral striate lesions in monkeys. Brain Res 120:209–220

Bruce C, Desimone R, Gross CG (1981) Visual properties of neurons in a polysensory area in superior temporal sulcus of the macaque. J Neurophysiol 46:369–384

Chen R, Classen J, Gerloff C, Celnik P, Wassermann EM, Hallett M, Cohen LG (1997) Depression of motor cortex excitability by low-frequency transcranial magnetic stimulation. Neurology 48:1398–1403

Cohen LG, Bandinelli S, Sato S, Kufta C, Hallett M (1991) Attenuation in detection of somatosensory stimuli by transcranial magnetic stimulation. Electroencephalogr Clin Neurophysiol 81:366–376

Cohen LG, Celnik P, Pascual-Leone A, Corwell B, Faiz L, Honda M, Dambrosia J, Sadato N, Hallett M (1997) Functional relevance of cross-modal plasticity in blind humans. Nature 389:180–183

Corbetta M, Miezin FM, Dobmeyer S, Shulman GL, Petersen SE (1990) Attentional modulation of neural processing of shape, color and velocity in humans. Science 248:1556–1559

Haxby JV, Grady CL, Horwitz B, Ungerleider LG, Mishkin M, Carson RE, Herscovitch P, Schapiro MB, Rapoport SI (1991) Dissociation of object and spatial visual processing pathways in human extrastriate cortex. Proc Natl Acad Sci USA 88:1621–1625

Hyvarinen J, Carlson Y, Hyvarinen L (1981) Early visual deprivation alters modality of neuronal responses in area 19 of monkey cortex. Neurosci Lett 26:239–243

Jenkins WM, Merzenich MM, Ochs MT, Allard T, Guic-Robles E (1990) Functional reorganization of primary somatosensory cortex in adult owl monkeys after behaviorally controlled tactile stimulation. J Neurophysiol 63:82–104

Kaas JH (1991) Plasticity of sensory and motor maps in adult mammals. Annu Rev Neurosci 14:137–167

Maccabee PJ, Amassian VE, Cracco RQ, Cracco JB, Rudell AP, Eberle LP, Zemon V (1991) Magnetic coil stimulation of human visual cortex: studies of perception. In: Magnetic motor stimulation: Basic principles and clinical experience (Levy WJ, Cracco RQ, Barker AT, Rothwell J, eds) Elsevier Science Publishers, B.V., Amsterdam, 111–120

Merzenich MM, Jenkins WM (1993) Reorganization of cortical representations of hand following alterations of skin inputs induced by nerve injury, skin island transfers, and experience. J Hand Therapy 6:89–104

Merzenich MM, Recanzone GH, Jenkins WM, Grajsk KA (1990) Adaptive mechanisms in cortical networks underlying cortical contributions to learning and nondeclarative memory. Cold Spring Harb Symp Quant Biol 55:873–887

Murata K, Cramer H, Bach-y-Rita P (1965) Neuronal convergence of noxious, acoustic and visual stimuli in the visual cortex of the cat. J Neurophysiol 28:1223–1239

Murray E, Mishkin M (1984) Relative contributions of SII and area 5 to tactile discrimination in monkeys. Behav Brain Res 11:67–83

Nudo RJ, Jenkins WM, Merzenich MM, et al. (1992) Neurophysiological correlates of hand preference in primary motor cortex of adult squirrel monkeys. J Neurosci 12:2918–2947

Pascual-Leone A, Torres F (1993) Sensorimotor cortex representation of the reading finger of Braille readers: An example of activity-induced cerebral plasticity in humans. Brain 116:39–52

Pascual-Leone A, Cammarota A, Wassermann EM, Brasil-Neto JP, Cohen LG, Hallett M (1993) Modulation of motor cortical outputs to the reading hand of Braille readers. Ann Neurol 34:33–37

Pascual-Leone A, Cohen LG, Brasil-Neto JP, Valls-Sole J, Hallett M (1994a) Differentiation of sensorimotor neuronal structures responsible for induction of motor evoked potentials, attenuation in detection of somatosensory stimuli, and induction of sensation of movement by mapping of optimal current directions. Electroencephalogr Clin Neurophysiol 93:230–236

Pascual-Leone A, Grafman J, Hallett M (1994b) Modulation of cortical motor output maps during development of implicit and explicit knowledge. Science 263:1287–1289

Pascual-Leone A, Grafman J, Hallett M (1994c) Transcranial magnetic stimulation in the study of human cognitive function. In: New horizons in Neuropsychology (Shugishita M, ed) Elsevier, Amsterdam, 93–100

Pascual-Leone A, Valls-Sole J, Wassermann EM, Hallett M (1994 d) Responses to rapid-rate transcranial magnetic stimulation of the human motor cortex. Brain 117:847–858

Pascual-Leone A, Alonso MD, Pascual-Leone Pascual A, Catalá MD (1995 a) Lasting beneficial effects of rapid-rate transcranial magnetic stimulation on slowness in Parkinson's disease (PD). Neurology 45 (suppl. 4):A 315

Pascual-Leone A, Dang N, Cohen LG, Brasil-Nets JP, Cammarota A, Hallett M (1995 b) Modulation of human cortical motor outputs during the acquisition of new fine motor skills. J Neurophysiol 74:1037–1045

Pascual-Leone A, Wassermann EM, Sadato N, Hallett M (1995 c) The role of reading activity on the modulation of motor cortical outputs to the reading hand in Braille readers. Ann Neurol 38:910–915

Pascual-Leone A, Rubio B, Pallardo F, Catála MD (1996 a) Rapid-rate transcranial magnetic stimulation of left dorsolateral prefrontal cortex in drug-resistant depression. Lancet 348:233–237

Pascual-Leone A, Tarazona F, Catala MD (1996 b) Modulation of motor cortical output maps associated with the acquisition of the Braille reading skill. Neuroimage 3:S 554

Pascual-Leone A, Grafman J, Cohen LG, Roth BJ, Hallett M (1997 a). Transcranial magnetic: a new tool for the study of higher cognitive functions in humans. In: Handbook of neuropsychology Vol. 11. (Grafman J, Boller F, eds) Elsevier B.V., Amsterdam, in press

Pascual-Leone A, Pujol J, Deus J, Capdevilla A, Tormos JM, Valls-Sole J (1997 b) Effects of repetitive transcranial magnetic stimulation on motor cortex activity during a rate controlled motor task as measured by functional magnetic resonance imaging. Neurology 48:A 106

Paus T, Jech R, Thompson CJ, et al. (1997) Transcranial magnetic stimulation during positron emission tomography: a new method for studying connectivity of the human cerebral cortex. J Neurosci 17:3178–3184

Rauschecker JP (1995) Compensatory plasticity and sensory substitution in the cerebral cortex. Trends Neurosci 18:36–43

Recanzone GH, Jenkins WM, Hradek GT, Merzenich MM (1992 a) Progressive improvement in discriminative abilities in adult owl monkeys performing a tactile frequency discrimination task. J Neurophysiol 67:1015–1030

Recanzone GH, Merzenich MM, Jenkins WM (1992 b) Frequency discrimination training engaging a restricted skin surface results in an emergence of a cutaneous response zone in cortical area 3a. J Neurophysiol 67:1057–1070

Recanzone GH, Merzenich MM, Jenkins WM, Grajski KA, Dinse HR (1992 c) Topographic reorganization of the hand representation in cortical area 3b owl monkey trained in a frequency-discrimination task. J Neurophysiol 67:1031–1056

Recanzone GH, Merzenich MM, Schreiner CE (1992 d) Changes in the distributed temporal response properties of SI cortical neurons reflect improvements in performance on a temporally based tactile discrimination task. J Neurophysiol 67:1071–1091

Sadato N, Pascual-Leone A, Grafman J, Ibanez V, Deiber MP, Dold G, Hallett M (1996) Activation of primary visual cortex by Braille reading in blind subjects. Nature 380:526–528

Shipp S, Zeki S (1989) The organization of connections between areas V5 and V1 in Macaque monkey visual cortex. Eur J Neurosci 1:309–332

Tergau F, Tormos JM, Paulus W, Pascual-Leone A, Ziemann U (1997) Effects of repetitive transcranial magnetic stimulation on cortico-spinal and cortico-cortical excitability. Neurology 48:A 107

Uhl F, Franzen P, Lindinger G, Lang W, Deecke L (1991) On the functionality of the visually deprived occipital cortex in early blind person. Neurosci Lett 124:256–259

Uhl F, Franzen P, Podreka I, Steiner M, Deecke L (1993) Increased regional cerebral blood flow in inferior occipital cortex and the cerebellum of early blind humans. Neurosci Lett 150:162–164

Ungerleider LG, Mishkin M (1982) Two cortical visual systems. In Analysis of visual behavior (Ingle DJ, Goodale MA, Mansfield RJW, eds) MIT Press, Cambridge, 549–586

The Perception of Actions:
Its Putative Effect on Neural Plasticity

J. Decety *

Summary

This chapter proposes that not only is bottom-up processing liable to affect cortical maps but top-down processing might also play a role. I will argue that the cognitive and neural mechanisms involved in the perception of human actions are worth investigating from a cognitive neuroscience perspective. I will marshal behavioral studies, physiological recordings in monkeys and positron emission tomography (PET) experiments in humans for a better and integrated understanding of the neural bases of perception of biological motion. I propose that top-down processing may have an effect on brain plasticity. Perception of action is also relevant to a contemporary theory of mind. Therefore, a final section deals with its implications for cognitive neuropsychiatry.

Introduction

Neural plasticity may be broadly defined as any change in the normal structure of the nervous system or its specific functions which is induced by injury or disease. One of the most widely accepted dogmas in neurology states that the nature of the mammalian brain is relatively stable after birth, or at best that sensory systems are highly plastic only during a short developmental time. This opinion has been challenged by a large body of studies that have demonstrated long-lasting changes in neural organization after brain injury (e.g., Ramachandran et al. 1992; Halligan et al. 1993). Functional rehabilitation following a lesion seems to include two processes that interact to maintain or restore an affected function. The first process, restitution, expresses the tendency of the neural network to recover itself when interrupted. This process seems to be the consequence of biochemical events in the nervous tissue. Hence it is relatively independent of the environment. The second process, substitution, implies a functional adaptation of the defective but partially restored network to compensate for the loss or disruption caused by the injury. This second process seems to depend strongly on external variables (sensory stimulation, physical therapy).

---

* Mental processes and brain activation – Inserm Unit 280, 151 cours Albert Thomas, 69424 Lyon cedex 03, France. And Cermep, 59 Bld Pinel, 69003 Lyon

J. Grafman / Y. Christen (Eds.)
Neuronal Plasticity:
Building a Bridge from the Laboratory to the Clinic
© Springer-Verlag Berlin Heidelberg New York 1999

In this chapter, I propose that not only is bottom-up processing liable to affect cortical maps (e.g. Mogilner et al. 1993; Xerri et al. 1994) but top-down processing might also play a role. My own contribution focuses on perception of action, whether it occurs from the environment (externally) or even when an individual mentally simulates an action (internally) that can be considered as an observation from the inside. This suggestion should be considered a working hypothesis since much work remains to be done. In addition, most of the data that will be discussed within a cognitive neuroscience framework are gathered from studies in normal subjects.

In the context of cognitive neuroscience, the issue of perception of action and its potential top-down effect on the neural machinery will be addressed in a section on psychology. Then electrophysiological recordings in awake monkeys, aimed at elucidating single cell properties in relation to the observation of action, are presented. This is followed by a section on human neuropsychology of the two major visual pathways, in which evidence from studies using positron emission tomography is presented. Actions may also be watched from the inside (motor imagery). This cognitive process, which may have a top-down effect on the activation of motor representation, is discussed. Finally, since perception of action is a central issue in the theory of mind, this chapter will end with a section on the relevance of this subject in psychopathology.

A Cognitive Neuroscience Perspective

The task of cognitive neuroscience is to map the information-processing structure of the human mind and to discover how this computational organization is implemented in the physical organization of the brain. Therefore, cognitive neuroscience is dealing with high cognitive processes such as categorization, mental imagery, attention, etc. Cognitive neuroscience is an interdisciplinary melding of studies of the brain, behavior and cognition, and of computational systems that have the properties of the brain and that can produce behavior and cognition (Kosslyn 1997). Not only does each approach constrain the others; rather each approach provides insights into different aspects of the same phenomena. In this heuristic perspective, information processing theory is not separate or independent of the properties of the neural substrate. This does not mean that the functional level is reducible to the physical mechanism, as wished by Churchland (1986), but that integration of top-down and bottom-up approaches is beneficial for the understanding of high level cognitive processes.

The fact that the human brain is an evolved biological system designed to solve a narrowly identifiable set of biological information-processing problems offers us the advantage of taking into consideration findings from non-human primates. However, marshalling facts from non-human primates and humans should be done with caution if one wants to avoid pitfalls in making inferences about the human brain from results of monkey studies. The human brain is the product of evolutionary processes acting over very long periods of time. While the human brain is both relatively large and exceedingly complex, it is similar in its organization to those of all other mammalian species, and it plays a similar role in the adaptation of the human species to its specific environmental niche (Tooby and Cosmides 1995). However, the evolution of the brain did not proceed by simple addition of parts to the brains of pre-existing species. The human brain is not that of an ancestral ape with the addition of a bit of frontal lobe and some language areas. In this view, the human mind represents not the highest expression of a common animal mind but rather one mind among many. The existence of neural diversity also suggests that nonhuman species have evolved cognitive specializations that are absent in humans (Preuss 1995).

Within the flow of external information vision plays a pre-eminent role among all sensory systems. From anatomy and physiology we know that in monkeys half of the neocortex is involved with the processing of visual information versus 20 to 30 % in human. From a psychological perspective, vision is the most often used not to say the most important system. Thus vision is the most represented sensory modality in the cerebral cortex, through which humans acquire most information about the surrounding world. And perception is the gateway to cognition.

Psychological Studies in Normal Subjects

A logical way to start this section is to inquire about the earliest developmental history of the normal child's understanding of the mind. It has been postulated that infants' primordial "like me" experiences are based on their understanding of bodily movement patterns and postures (Meltzoff and Gopnik 1993). Infants monitor their own body movements by the internal sense of proprioception and can detect cross-modal equivalents between those movements-as-felt and the movements they see performed by others. Early imitation in infants seems an intentional matching to the target provided by the other, rather than a rigidly organized purely reflexive response as a behaviorist model would suggest (Konorski 1967). The hypothesis suggested by Meltzoff and Moore (1992) is that imitation is based on infants' capacity to register equivalence between the body transformations they see and the body transformations they only feel themselves make.

Human movements are not perceived as mere changes in the location of body parts, but as walking, reaching for an object, running, limping, talking and so on. The Swedish psychologist Johansson (1973) demonstrated that the kinematic pattern was sufficient to allow perception of persons and what they were doing. His studies were based on the point-light technique, which used small light-reflecting patches attached to the main joints (e. g., wrists, knees, ankles) of actors dressed entirely in black so that only the lights were visible. Then the actors were filmed at they moved around. This technique excludes potential influences of a priori knowledge on perception and thus isolates the role of visual cues. Static presentation of the point-light stimulus did not help in identification

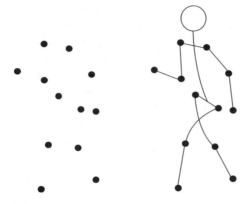

Fig. 1. Static illustration of the point-light technique used by Johanson (1973). Lights attached to a person's joints are not perceived as a recognizable object in the dark when the person stands still. When the person begins to move, the lights are perceived immediately as a human form.

or in recognition of the stimulus. However, as soon as the stimulus moved, subjects could recognize or discriminate within 200 ms the movement of the actors (Fig. 1) as far as the orientation in the frontoparallel plane is concerned. Using the same basic paradigm, Kozlowski and Cutting (1977) extended these findings by showing that observes can make very precise discriminations when watching point-light displays, such as the recognition of the gender of walkers. Even more remarkable is the fact that observes can identify themselves and others known to them (Cutting and Kozlowski 1977). However, when the configuration was presented upside down, observers did not report seeing an upside-down human figure (Sumi 1984). In a series of long-term priming experiments, Verfaillie (1993) provided evidence that recognition of human biological motion is viewpoint specific, which suggests that recognition is accomplished by accessing high-level, orientation representations. In this context, priming is based on a reactivation of the representations mediating target identification. The fact that non-human primates and humans think in categories and that in the latter, concepts influence the way of thinking and acting has lead Dittrich (1993) to conduct a study to test whether the ability to detect natural motions is in part determined by the content of independent categories of the information that physically characterizes the event. In his study, locomotory, instrumental and social actions were presented with the point-light display technique in a normal situation (light attached to joints), inter-joint (light attached between joints) and upside-down. Subjects' verbal responses and recognition times showed that locomotory actions were recognized much better and faster than social and instrumental actions. Furthermore, biological motions were recognized much better and faster when the light-spot displays were presented in the normal orientation rather than upside down. Finally, recognition rate was only slightly impaired under the inter-joint condition (Fig. 2).

These data suggest that the perceptual analysis of actions and movements starts primarily on an intermediate level of action coding and comprises more than just the similarity of movement patterns or simple structures. It can be also argued that coding of dynamic phase relations and semantic coding take place at very early stages of the processing of biological motion. From an evolutionary

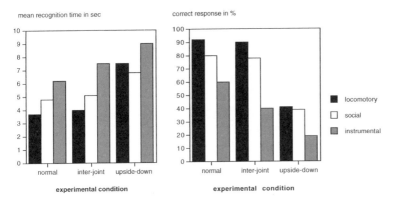

Fig. 2. Mean recognition time (sec) on the left and recognition accuracy (%) on the right for each of three types of movements in three experimental conditions: normal (the light-spots were attached at the joints), inter-joint (midway between the joints) and upside-down (light-spots attached at the joints but the actor was filmed upside-down). Averaged data from 28 subjects. Adapted from Dittrich 1993 with permission.

standpoint, it is reasonable to expect that, among the wide range of goal-directed actions, humans recognize faster those biological motions such as locomotion which were also detected by our ancestors. Conversely, instrumental actions are specific to humans and social actions are in between.

Another interpretation of the recognition of biological motion is that perception and recognition processes are mediated by the implicit knowledge of production (motor) rules. This idea is supported by experiments in the domain of handwriting. Viviani and Stucchi (1989) have shown that the visual perception of a simple geometrical figure is influenced by implicit knowledge about the laws of graphic production. Several authors have suggested that motor knowledge can be used to anticipate forthcoming sequences when perceiving human movements (Shiffrar and Freyd 1993; Orliaguet et al. 1998). The ability to perceive biological motion does not seem to require complex cognitive processes, since young children are capable of detecting such stimuli and are able to make predictions with no apparent difficulty (von Hofsten 1993).

Perception is known to be closely linked to action. It can no longer be a simple process of passive transformation from sensory transducer activity into central representations (Berthoz 1996). Indeed the control of action requires predictive mechanisms which in turn require a preselection of relevant sensory information. Perception thus serves to predict the consequences of actions but also the intentionality of the observed behavior. In an experiment by Runeson and Frykholm (1983), actors were asked to lift a box weighing four kilograms and to carry it to a table, while trying to give the impression that the box weighed 6.5, 11.5, or 19 kilograms. Observers detected the actors' intentions from the pattern of lights, and thus were not deceived about the actual weight of the box.

Neurophysiological Recordings in Primates

Two brain regions in the monkey have been found to be implicated in the perception of body movements. These regions concern specific cortical areas in the temporal cortex and in the frontal lobes. Extensive chronic single-unit recording in different regions of the temporal association cortex have been performed by the group of David Perrett at the University of St. Andrew. These authors have discovered populations of cells in the upper bank of the superior temporal sulcus (STS) which appear to be involved in the recognition of individuals and how these individuals are moving (Perrett et al. 1987). In an adjacent section of STS, in the lower bank, populations of neurons were found to be responsive to the sight of action, that is, how other individuals are interacting with the environment (Perrett et al. 1989). The same group found cells specifically responsive to hand-object interactions (Perrett et al. 1990). The cells studied were not equally responsive to the observation of all hand actions. Some cells were highly selective for different actions such as reaching for, retrieving, manipulating, picking, tearing, holding. A clear selectivity for the agent performing an action was found for hand-object interactions over object-object interactions, despite similar eye movements accompanying both actions. According to the authors, the finding of cells that are selective for the sight of actions and that are unaffected by auditory cues associated with actions indicates the extent to which meaningful relations can be derived purely within the reliance on the capacity for language. These data provide an opportunity for direct study of the neuronal mechanisms by which the brain computes interactions and determines causal and intentional relation within action.

The second region found to be involved in the observation of action has been identified in the rostral part of inferior area 6. This agranular frontal region, also termed as area F5, receives input from the inferior parietal lobule. It was recently discovered that a particular subset of neurons in this frontal region discharge both during a monkeys active movements and when the monkey observes meaningful hand movements made by an experimenter or by another animal (Di Pellegrino et al. 1992). The experimenter's movements included, among others, placing or taking away objects from a table, grasping food, manipulating objects. Thus while these neurons have similar visual properties to those of STS neurons, they differ from them in the fact that they discharge also during active movements. These neurons, named "mirror" neurons, suggest that this region is endowed with an observation/execution matching system. Rizzolatti and coworkers (1996a, b) proposed that this observation/execution mechanism plays a role in understanding the meaning of actions. Based on this finding and given the proposed homology between F5 and human Broca's region, it has been posited that a similar matching system exists in humans and that it could be involved in recognition of actions as well as phonetic gestures (Gallese et al. 1996).

Another brain region involved in the visual analysis and control of action lies in the parietal cortex. Recent neurophysiological studies in alert monkeys have revealed that the parietal association cortex plays a crucial role in depth perception and in visually guided hand movement. Sakata et al. (1997) suggested that neural representation of 3-D objects with real physical dimensions and their ego-centric positions and movements seems to occur in the parietal association cor-tex. The major purpose of the 3-D representation is the visual guidance of goal-directed action. As noted by Carrey et al. (1997), it is not clear whether cells within the parietal cortex code the sight of actions executed by other individuals or utilize such information to enable imitation or other types of interaction.

Last but not least is the issue of the integration and maintenance of behavior-ally relevant information, in which the prefrontal cortex makes a major contribu-tion. This region receives inputs from virtually all brain sensory systems and notably from the ventral and the dorsal pathways. Rao et al. (1997) have exam-ined the activity of a large group of cells within the primate prefrontal cortex during a task that engaged both "what" and "where" working memory. Some neurons showed either object-tuned (what) or location-tuned (where) delay activity. However, and most importantly, over half of a percent of these neurons with delay activity showed both what and where tuning and were interpreted as contributing to the linking of object information with spatial information needed to guide behavior. Another similar experiment suggested that the prefrontal cor-tex not only plays a primary role in working memory but also be a source of top-down inputs to the ventral pathway, biasing activity in favor of relevant stimuli (Miller et al. 1996).

Neuropsychology in Humans

The visual system has been able to accommodate two somewhat distinct func-tions, one concerned with representing the world and the other with acting on it. In evolutionary terms, the earliest functions of vision were action- rather than perception-oriented (Milner and Goodale 1995). These two distinct functions appear to depend on separate neural mechanisms. Most of the work concerning the identification of the wiring subserving this division of labor in the visual sys-tem is based on non-human primates. Ungerleider and Mishkin (1982) were able to distinguish between two main streams of projections arising from the primary visual cortex and projecting to higher visual areas: the ventral stream of visual processing projects to the inferotemporal cortex, while the dorsal stream projects instead to the parietal cortex. The dorsal stream plays a key role in object identi-fication and the dorsal stream is responsible for localizing objects in visual space. Goodale proposed (1993, 1997) that both streams process information about object features and about their spatial relations, but each stream uses this visual information in different ways. In the ventral stream, the transformations focus on the stable characteristics of objects and their relations, allowing the formation of long-term perceptual representations. These representations play an essential role in object recognition and categorization. In contrast, the transformations carried out by the dorsal stream deal with the moment-to-moment information about location and spatial disposition of objects in particular egocentric coordi-nates and thereby mediate the visual control of skilled actions.

The model postulated by Goodale is grounded in the observation of selective dissociation between perception and action in neurological patients, such as the case of DF who demonstrated accurate guidance of hand and finger movements directed at the very objects whose qualities she failed to perceive (Goodale et al. 1991). A reverse dissociation has been reported by Jeannerod et al. (1994). Several PET studies performed in normal subjects using tasks involving recognition of visual patterns and object localization support the dual pathway in the ventral stream whereas activation was located in the dorsal stream during spatial matching tasks (Haxby et al. 1991; Köhler et al. 1995). We have recently scanned subjects during an object-matching task and a grasping task of the same objects. Our results support the idea that spatial vision and object identification are mediated by two distinct systems. However, object-oriented action and object matching activate a common posterior parietal area, namely in the right intraparietal sulcus, suggesting that object analysis based on geometrical cues is processed by this area, whatever the goal of the task (Faillenot et al. 1997).

A more direct approach to the search of neural regions selectively responsive to the observation of action relies on brain mapping studies that specifically address the neural correlates of perception of biological motion. Relatively few PET studies have been performed so far but their results are promising. The first study was conducted by Bonda et al. (1996). Healthy subjects were scanned during the perception of stimulations of biological motion with the point-light display, such as the one elaborated by Johansson (1973). Four experimental conditions were used: perception of goal-directed hand movements (i.e., grasping a glass and bringing it to the mouth), whole body movements (i.e., moving backward and forward and from left to right), object motion (abstract geometrical stimuli) and random motion patterns. Their results demonstrated that perception of goal-directed hand action implicated the anterior part of the intraparietal sulcus (Brodmann area (Ba)40) and the caudal part of the superior temporal sulcus (Ba 39), both in the left hemisphere. By contrast, perception of whole body movements was associated with activations of the rostrocaudal part of the right superior temporal sulcus (Ba 21/37) and adjacent temporal cortex, and the amygdala bilaterally. According to the authors, these results demonstrate that the interpretation of different types of biological movements (hand actions and whole body movements) engages different brain areas. Perception of goal-directed hand action selectively activated regions in the posterior part of the parietal cortex and the superior temporal sulcus, in accordance with the model of Goodale (1993). In contrast, perception of whole body movements involves interaction between temporal neocortex and limbic system, which is critical for emotional effector patterns of behavior.

In a study performed by Rizzolatti et al. (1996b), human subjects were scanned under three conditions: observation of grasping common objects, grasping the same objects and object observation as a reference task, an experimental paradigm close to that used in monkeys by the same authors. The results of the observation of grasping objects from which the observation of the same objects was subtracted indicate regional cerebral blood flow (rCBF) activations in the left middle temporal gyrus including that of adjacent superior temporal sulcus, corresponding to Brodmann area 21, and in the caudal part of the left inferior frontal gyrus (Ba 45). Their data indicate that there is good agreement between the functional areas in the human brain and in the monkey brain devoted to hand action recognition. The activation in the left temporal lobe would correspond to the STS in monkeys (Perrett et al. 1989) and the activation in the pars triangularis (Ba 45) might be similar to F5.

To explore the effect of the semantic content of biological motion as well as the effect of the cognitive strategy during observation, Decety et al. (1997) measured cerebral metabolic activity in healthy volunteers while they were engaged in the observation of pre-recorded meaningful and meaningless hand movements on a video monitor. The meaningful movements consisted of pantomimes, e.g., opening a bottle, drawing a line, sewing a button, hammering a nail, etc. These movements mainly involved the right (dominant) hand; the left hand was used to hold the imaginary object. Meaningless movements were derived from American Sign Language (ASL), with the constraints that they should be physically and perceptually as close as possible to the actions presented during the meaningful actions (e.g., movements involving mainly the right hand). As the subjects were unacquainted with ASL, they were unable to relate such actions to language or symbolic gestures. For both types of actions (meaningful and meaningless), subjects were instructed to observe the videos carefully with one of two purposes: either to "imitate" or to "recognize" the actions after the scanner acquisition.

It was found that the meaning of the movements, irrespective of the strategy used during observation, lead to different patterns of brain activity and clear left/right asymmetries (Figs. 3 and 4).

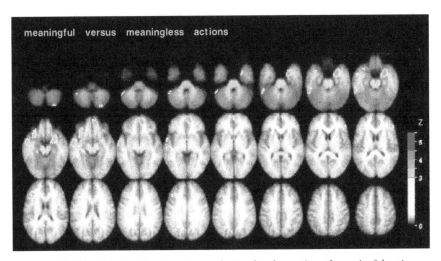

Fig. 3. Cerebral localization of activated areas during the observation of meaningful actions versus meaningless actions, irrespective of the strategy. PET data are superimposed on an averaged MRI (transverse sections parallel to the AC-PC plane). Left hemisphere is on the left

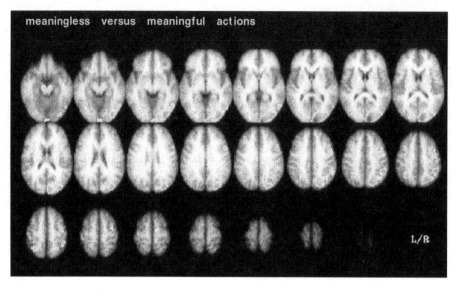

Fig. 4. Cerebral localization of activated areas during the observation of meaningless actions versus meaningful actions, irrespective of the strategy. PET data are superimposed on an averaged MRI (transverse sections parallel to the AC-PC plane)

Meaningful movements relative to meaningless movements, irrespective of the strategies, strongly engaged the left hemisphere in frontal regions including the inferior frontal gyrus (Ba 45), the orbital cortex (Ba 11 and 47), and the middle temporal gyrus (Ba 21). Meaningless movements relative to meaningful movements involved mainly the right occipito-parietal pathway (Ba 18, 19, 7 and 40), extending to the premotor cortex (Ba 6). Observing with the intent to imitate, irrespective of the content (meaningful or meaningless) of the presented movements, was associated with bilateral activations of the dorsolateral prefrontal cortex, the anterior SMA and the cerebellum on the left side (Fig. 5).

In contrast, observation with the intent to recognize activated the right parahippocampal gyrus and the left insula. The results obtained in this study during action observation clearly demonstrate that the two types of action lead to different neural networks. Indeed, the patterns of cortical activation corresponding to these two types differ both in terms of hemispheric asymmetry and repartition of the involved areas. The network activated during observation of meaningful actions in the left hemisphere corresponds to the "ventral" visual pathway, which includes inferotemporal areas and part of the hippocampus and terminates in the inferior frontal gyrus. On the other hand, the network associated with meaningless actions in the right hemisphere corresponds mainly to the "dorsal" pathway, which includes occipitoparietal areas and is connected with premotor cortex.

The subjects' aim during observation of actions and movements elicited different patterns of brain activation. When the aim of observation is to imitate, the structures that are usually involved in the planning of action were found to be

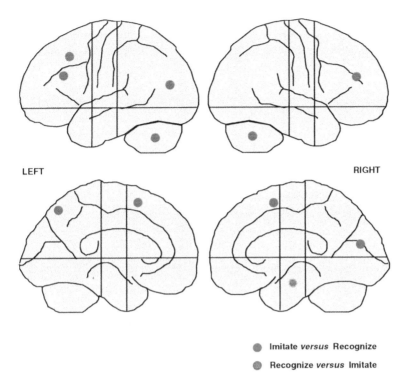

LEFT   **RIGHT**

● Imitate *versus* Recognize
● Recognize *versus* Imitate

Fig. 5. Cerebral localization of activated areas during the observation of actions, elicited by subject's strategy (intention to recognize and to imitate), irrespective of the nature of the stimuli (meaningful and meaningless actions). Lateral views are shown on the top, and medial views on the bottom of the figure. The horizontal line indicates the anterior-posterior commissure (AC-PC). The vertical planes pass through the anterior (VAC) and through the posterior (VPC) commissures. Adapted from Decety et al. (1997)

activated: the dorsolateral prefrontal cortex, the anterior SMA and the lateral cerebellum. These structures are engaged in the generation and selection of action (Passingham 1996).

Thus the brain regions that are active during observation of actions are dependent, in part, on the nature of the required executive processing (imitation/recognition) and are strongly modulated by the type of the intrinsic properties of the movement presented. These mechanisms might represent the neural basis for imitation, observational learning and motor imagery. Finally, these inferior frontal and middle temporal regions appear to be selectively responsive to biological stimuli. Indeed, a PET experiment in which subjects were presented 3-D objects grasped by an image of a hand in a virtual reality system failed to activate these two regions (Decety et al. 1994). The observation of the virtual hand grasping virtual objects mainly activated visual areas and especially those involved in motion detection. The virtual technology was thus unable to create a hand that was perceptually identical to a natural one.

Activation could have been expected in the posterior and inferior parietal cortex, since this region is known to play a key role in action representation (De Renzi 1989; Rothi et al. 1991; Sirigu et al. 1995, 1996). The lack of detection of this cortical region in this experiment may be a consequence of the subtractions between conditions that all included hand movements and /or the absence of a right baseline state.

To further investigate the respective effect of meaningful and meaningless actions, as well as the potential contribution of the aim of the observation on the neural network engaged by the perception of biological motion, we have performed a second PET study (Grèzes et al., 1998 in press) on another group of subjects. The stimuli were the same as those described above. The experiment was conducted in two sessions. In the first one, subjects were requested to watch meaningful and meaningless actions without any specific purpose. In the second session, subjects were instructed to watch these actions with the aim of imitating them immediately after the end of the scanning acquisition. A reference condition was also included in which stationary hands were presented. The stimulus structure was visually the very same as that used in activation tasks with the following characteristic: no movements. Five spatial positions of the hands and limbs were used and randomly presented throughout the condition. The aim of this condition was to provide a reference level for activation tasks, i. e., to subtract the "low level" visual analysis of the stimuli (upper limb recognition, color texture) and thus demonstrate the wole neural response to the perception of biological motion.

Perception of meaningful movements without any purpose versus static hands was associated with rCBF increases in the superior occipital gyrus (Ba 19) and in the occipital temporal junction (Ba 19/37) bilaterally. The inferior frontal gyrus (Ba 44/45), the insula, the middle temporal gyrus (Ba 21) and the fusiform (Ba 20 and Ba 37) and the inferior parietal lobe in its lower part (Ba 40) were activated in the left hemisphere, whereas the lingual gyrus (19/18) was activated on the right side.

Activations elicited by the perception of meaningless movements without any purpose versus static hands were located in the superior occipital gyrus (Ba 19) and the occipital temporal junction (Ba 19/37) bilaterally. In addition rCBF increases were found in the superior parietal lobule (Ba 7), the intraparietal sulcus (7/40), inferior parietal lobe in its lower part (Ba 40) and the middle temporal gyrus (Ba 21) in the left hemisphere. Activations were also found in the superior parietal lobule, the inferior parietal lobe in its upper part (Ba 40) and in the cerebellum on the right side.

In summary, perceptions of hand actions, whether they are meaningful or meaningless, share, to some extent, a common cortical network. This network consists of the superior occipital gyrus (Ba 19), the occipital temporal junction

(Ba 19/37) in both hemispheres, the middle temporal gyrus (Ba 21) and the inferior parietal lobe in its lower part (Ba 40) in the left hemisphere. One may suggest that these activations are related to the analysis of hand movement. The activation of the occipito-temporal junction (Ba 19/37) corresponds precisely to the coordinates of V5 given by Watson et al. (1993) from a PET study in humans. This region is known to be specifically engaged by motion perception. The activation of Brodmann area 21 is located in the posterior part of the ventral stream, in which visual information is processed following a gradient, from simple to complex attributes (Ungerleider 1995). This region could be involved in the fine analysis of hand-finger movements and their relation to the object-use. In addition to this common network, meaningful and meaningless actions engaged the following specific brain regions:

– meaningful actions activated the precentral gyrus (Ba 6/4), the inferior frontal gyrus (Ba 44/45) and the fusiform gyrus (Ba 37/20) in the left hemisphere. On the right side, the lingual gyrus (Ba 19/18) was activated.

– meaningless actions activated the superior parietal lobule (Ba 7) in both hemispheres, the inferior parietal lobe in its upper part (Ba 40) and the cerebellum in the right hemisphere.

These results support the PET data from the previous experiment. However, the activation of the dorsal pathway is now bilateral, not restricted to the right hemisphere as in the first study, although it is stronger on the right. This finding is likely due to the control condition (static hands), which was subtracted from this condition.

The effect of the strategy (i.e., intention to imitate) confirmed and also extended the results of the first study. Observation in order to imitate versus observation without any purpose, irrespective of the nature of the stimuli, was associated with activations located bilaterally in the middle frontal gyrus (Ba 9), in the dorsal premotor cortex (Ba 6) and in the SMA. The inferior parietal gyrus (Ba 40), in its upper part, the superior parietal lobule (Ba 7) and the anterior cingulate gyrus (Ba 32) were activated in both sides. At the subcortial level, the mediodorsal thalamic nucleus was found to be activated in the right hemisphere, whereas the posterior caudate nucleus was activated in both sides. The dorsal frontal gyrus (Ba 10), the cerebellum and the cuneus (Ba 17/18) were activated in both hemispheres. rCBF increases were also found in the precuneus (Ba 7) and in the middle frontal gyrus (Ba 46) in the left hemisphere. The whole set of brain regions involved in motor representation is activated during the observation stage when the goal is to reproduce the actions perceived.

These findings support psychological evidence that observation of action provides an effective means of learning new skills (Carroll and Bandura 1990). This capability points towards a close linkage between representations of events in the environment and representations for action (Vogt 1996).

Neuroimaging investigations provide a unique opportunity to localize and define the neural substrate of perception of biological motion in humans. Fortunately, the results are coherent with electrophysiological studies performed in monkeys. Several distributed cortical areas (frontal, temporal and parietal) contribute to the recognition of action. It is clear that the neural substrates for producing actions are intimately associated with the neural machinery responsible for perceiving actions (Carey et al. 1997). Another argument in favor of the close linkage between perception and action is provided by the experiment conducted by Fadiga et al. (1995), in which subjects were asked in one condition to observe grasping movements performed by an experimenter as compared with appropriate control conditions. At the end of the observation period, a transcranial magnetic stimulus was applied over their motor cortex while motor evoked potentials were recorded from hand muscles. It was found that the pattern of muscular response was selectively increased with respect to control conditions. Furthermore, the pattern reflected the same as that recorded during actual grasping movement.

Observation of Action from the Inside

One of perhaps the most evolved mechanims in the human brain is our ability to simulate our behavior. When mental simulation is directed on a given action with the subject experiencing from within, it is termed motor imagery. It has been suggested that motor imagery should be considered as a virtual simulation of motor behavior (for a review, Decety 1996a). Such an "inner" simulation process, experienced consciously, requires the construction of a dynamic representation in working memory that makes use of spatial and kinesthetic components retrieved from long-term memory, as well as the activation of serial plans of action. Such processes apparently require internal intracerebral feedback loops that involve subcortical structures (Decety and Ingvar 1990). Motor imagery in normal and brain-damaged subjects may be used for deciphering the content and structure of covert processes that precede the execution of action (Jeannerod and Decety 1995). Broadly defined, motor representations account for the cognitive mechanism related to the goal and the consequences of an action, whether it is internally or externally triggered (Decety 1996b for a review). Several PET studies have demonstrated that motor imagery is associated with activations in the SMA, in the inferior parietal lobule, premotor cortex, dorsolateral prefrontal cortex, anterior cingulate and cerebellum. At the subcortical level, activations were observed in the basal ganglia (Roland et al. 1980; Decety et al. 1994; Lang et al. 1994; Stephan et al. 1995; Parsons et al. 1995). Functional magnetic resonance imaging experiments have recently confirmed these findings, but in addition they have reported the involvement of the primary motor cortex (M1) during motor imagery (Leonardo et al. 1995; Roth et al. 1996). The fact that not only areas upstream to M1 but also M1 itself is activated during mental simulation, is a strong argument for the hypothesis that the neural mechanisms involved in the mental representation of an action are common to those used for its execution.

The neural substrate of motor representations is largely distributed within the central nervous system, in which the dorsolateral prefrontal cortex plays a major executive role in the generation and supervision of action (Grafman 1989).

Another key region is the parietal cortex, whose role in the ability to generate movement representations is supported by results from a wide range of studies in electrophysiological recording in monkeys (Duhamel et al. 1992; Kalaska and Crammond 1995), and clinical observation in apraxic patients (Sirigu et al. 1996; Rothi and Heilman 1997).

Recently, Crammond (1997) suggested that the parietal cortex is a structure where the evaluation of motor performance, real and also intended, can occur based upon a comparison of corollary discharge and multimodal reafferent sensory signals with a stored internal representation of the motor plan as it normally unfolds. Specific disruptions of separate subregions with the posterior parietal cortex may produce a wide range of deficits related to action representation (action imitation, action recognition and action generation). There are also deficits in visual comprehension of action that can be independent of motor production (Rothi et al. 1986). Those deficits are frequently observed in ideomotor apraxis when the patient is asked to pantomime symbolic gestures, to imitate, etc. However, these cognitive deficits do not arise when the actions are embedded in a natural context (Heilman et al. 1997).

Thus, the top-down activation of motor representations would involve virtually all stages of motor control.

Such an hypothesis has strong implication for the putative effect of motor imagery on brain plasticity. The influence of mental training on motor skill learning has been acknowledged for quite a long time (Feltz and Landers 1983). But it is only recently that neurophysiological evidence has been reported to explain this effect. Yue and Cole demonstrate (1992) that strength increases may result from mental training-induced changes in voluntary motor programs. The authors compared, in normal subjects, the maximal voluntary force production of the fifth digit's meta-carpophalangeal joint after a training program of repetitive maximal isometric muscle contractions and after a motor imagery training that did not involve repetitive activation of muscle. The average abduction force of the left (trained) digit increased 22 % for the imagining group and 30 % for the contraction group. The maximal abduction force of the right (untrained) fifth digit increased significantly in both the imaging and contraction groups after training. Thus, strength increases can be achieved without repeated muscle activation. These force gains appear to result from practice effects on central motor programming.

This top-down effect has also been demonstrated by Pascual-Leone et al. (1995) in a study of plastic changes of the human motor cortex in the acquisition of new fine motor skills with the use of transcranial magnetic stimulation (TMS). These authors mapped the cortical motor areas targeting the contralateral long finger flexor and extensor muscles in subjects learning a one-handed, five-finger exercise on the piano. In a second experiment, the different effects of mental and physical practice of the same five-finger exercise on the modulation of the cortical motor areas targeting muscles involved in the task were measured. The results showed that mental practice led to the same plastic changes in the motor system as those occurring with the acquisition of the skill by repeated physical practice.

Thus, mental practice seems sufficient to promote the modulation of neural circuits involved in the early stages of motor learning. The authors proposed that this modulation may occur through an increase of synaptic efficacy in existing neural circuits (long-term potentiation) or unmasking of existing connections due to disinhibition.

If one accepts the fact that neural plasticity may be induced by top-down processing, such as observation of action or mental simulation, then an important application could be mode in brain-damaged patients to contribute in recovery from motor deficits.

Functional magnetic resonance imaging (fMRI) has the potential to disclose the neural processes underlying the cognitive mechanisms subserving perception of action, motor representations and the coupling between them that can determine the interpretation and the understanding of intentions. Although the temporal window offered by fMRI is ultimately limited by the physiological processes (hemodynamic responses takes over seconds), it is nevertheless much better than PET. Furthermore, methods for improving the temporal resolution are currently under construction (Appolino et al. 1997).

The Relevance of Perception of Action to Psychopathology

Understanding the way other's people's mind work, and knowing how those minds are similar to or different from our own mind, is crucial if one wants to interact with people. Although we directly observe other people's behavior, we think of them as having internal mental states that are analogous to our own. We think that human beings want, think, and feel, and that these states lead to their actions (Meltzoff and Gopnik 1993). Within the same psychological context, it is now admitted that thoughts, beliefs and ideas are not only distinct from the physical world of objects and behavior, but are also causally related to that physical-behavioral world. Causal influence goes from mind to world and from world to mind: mental states cause actions in the world and the world causes mental states (Wellman 1993).

The concept of theory of mind was coined by Premack and Woodruff (1978) in their work on primates. Leslie (1987) has developed a theoretical analysis of the representational mechanism underlying this ability in children who, by the age of three or four, are able to understand mental states. This theory was then used to describe the deficit presumed essential to the pathology of autism (Baron-Cohen 1990). It is indeed frequently reported that people with autism tend to lack theory of mind. (For a review of the theory of mind in autism see Frith 1989.) One striking feature that has social, communicative and cognitive implications is that autistic children are deficient in imitating the actions of others. Imitative ability is the mechanism by which normal children develop an understanding of minds (Meltzoff and Gopnik 1993). Through imitation, the child perceives others as similar to self and learns to detect cross-modal equivalence between others' action (as seen) and his own internal states (as perceived).

Therefore, deficiencies in the ability to imitate others' actions may contribute to the failure to develop a normal theory of mind (Barresi and Moore 1996). Smith and Bryson (1994) have suggested that, before appealing to a deficit at the conceptual level, a more fundamental inquiry into the representation of the movement components of actions and their associated functional representations should be considered. According to these authors, low-level deficits may, depending on their developmental timing and severity, have devastating impacts on higher level processes. Accordingly, abnormality of executive functioning may derive ultimately from atypicalities in perceiving and representating actions and objects in the world. Smith and Bryson concluded their critical review paper by pleading for future research that should address the specific motor and praxis functions that are intact in some people with autism and impaired in others. They also suggest, among several directions, the use of a more analytic approach to the praxic problems, such as the one in neuropsychology of apraxic patients dealing with the generation, access or manipulation of motor representations.

Nowadays, theory of mind has become the focus of increasing attention, not only in psychology but also in cognitive neuroscience. Baron-Cohen (1994) has proposed a cognitive modular system that enables attribution about mental states of others. This system relies on four independent modules: an intentionality detector, an eye direction detector, a shared attention mechanism and a theory of mind mechanism. Perrett and Emery (1994) reviewed the properties of cells in the monkey temporal cortex that are selectively activated by visual cues of other individuals and that might correspond to each of the subcomponents of the model hypothesized by Baron-Cohen (1994), with the exception of the latter component for which there is no neurophysiological evidence. Apparently, this is not the case in humans. For instance, Goel et al. (1995) have reported a PET activation study performed in normal volunteers while performing tasks involving the attribution of intention to others. In their experimental conditions, subjects were exposed to familiar and unfamiliar objects and they were required to make distinctive decisions. In the theory of mind condition tasks, they were requested to judge whether a 15th century European would be familiar with the function of a particular object. The main finding of this study was the activation of the left medial prefrontal cortex (Ba 9) and of the left temporal lobe (Ba 21, 39/19, 38) during the theory of mind tasks. In another study reported by Fletcher et al. (1995), normal subjects performed story comprehension tasks necessitating the attribution of mental states. Compared to control tasks that did not require mental attribution, theory of mind stories involved a specific pattern of activation in the left medial prefrontal cortex, on the border between Brodmann areas 8 and 9 and in the posterior cingulate gyrus. Lack of a selective implication of the left medial prefrontal cortex in understanding other minds has been reported in a recent PET study by Happé et al. (1996), performed with patients diagnosted with Asperger syndrome, i.e., autistic subjects who tend to be older and more verbal than other autistic subjects and who show success across an array of test that involve mentalizing. The Asperger patients showed activation of an adjacent, but more ventral area of the medial prefrontal cortex (Ba 9/10). According to the

authors, this difference suggests that the Asperger subjects mentalizing performance was subserved by a brain system in which one key component was missing.
The key component, at the cognitive level, is the capability to integrate information in context to extract higher level meaning or gist. The neural substrate for
this key component is the left medial prefrontal cortex.

Thus, these studies conclude that they have discovered the neural substrate of
attribution of intention, namely a rather well-circumscribed cortical area within
the medial frontal cortex, in the left hemisphere. Yet, this result and reasoning
may appear odd to many cognitive psychologists as well as to neuroscientists, in
the sense that it clearly postulates that a high level and rather central information
processing mechanism (in Fodor's sense, which would correspond to a horizontal
faculty, 1983) is grounded on a single cortical region. Such a strong inference is
puzzling when, at the same time, most neuroimaging studies demonstrate that
many functions depend on complex interactions between distant and distributed
brain areas. However, one should keep in mind that localization from neuroimaging techniques is based on thresholding the signal and thus only the peaks of the
iceberg are visible.

Acknowledgments

The writing of this paper was supported by the Inserm, and by grants from
Biomed 2, GIS-CNRS and the Fondation pour la Recherche Médicale. Thanks are
due to Dr. Jordan Grafman (NIH, Bethesda) for helpful comments on an earlier
draft of the paper.

References

Appolino I, Rueckert L, Partiot A, Turner R, Le Bihan D, Jezzard P, Grafman J (1997) Functional mag-
    netic resonance imaging: basic principles and applications in neuropsychology. In: Boller F, Grafman
    J (eds) Handbook of neuropsychology. Vol 11. Action and Cognition. Elsevier, Amsterdam, pp
    211–266
Baron-Cohen S (1990) A specific cognitive disorder of "mind-blindness". Int Rev Psychiat 2: 81–90
Baron-Cohen S (1994) How to build a baby that can read minds: cognitive mechanisms in mind read-
    ing. Cahiers Psychol Cogn 13: 513–552
Barresi J, Moore C (1996) Intentional relations and social understanding. Behav Brain Sci 19: 107–154
Berthoz A (1996) Neural basis of decision in perception and in the control of movement. In: Damasio
    AR, Damasio H, Christen Y (eds) Neurobiology of decision making. Springer Verlag, Berlin, pp
    83–100
Bonda E, Petrides M, Ostry D, Evans A (1996) Specific involvement of human parietal systems and the
    amygdala in the perception of biological motion. J Neurosci 16: 3737–3744
Carey DP, Perrett DI, Oram MW (1997) Recognizing, understanding and reproducing action. In: Bolle
    F, Grafman J (eds) Handbook of neuropsychology. Vol 11. Action and Cognition. Elsevier, Amster-
    dam, pp 111–129
Carroll WR, Bandura A (1990) Representational guidance of action production in observational learn-
    ing: a causal analysis. J Motor Behav 22: 85–97

Churchland PS (1986) Neurophilosophy. Toward a unified science of the mind-brain. Cambridge, MA, MTI Press

Crammond DJ (1997) Motor imagery: never in your wildest dream. Trends Neurosci 20: 54–57

Cutting JE, Kozlowski LT (1977) Recognising friends by their walk: gait perception without familiarity cues. Bull Psychonomic Soc 9: 353–356

Decety J (1996a) Do imagined and executed actions share the same neural substrate? Cogn Brain Res 3: 87–93

Decety J (1996b) Neural representations for action. Neurosci 7: 285–297

Dectey J, Ingvar DH (1990) Brain structures participating in mental simulation of motor behavior: a neuropsychological interpretation. Acta Psychol 73: 13–34

Decety J, Perani D, Jeannerod M, Bettinardi V, Tadary B, Woods R, Mazziotta JC, Fazio F (1994) Mapping motor representations with positron emission tomography. Nature 371: 600–602

Decety J, Grèzes J, Costes N, Perani D, Jeannerod M, Procyk E, Grassi F, Fazio F (1997) Brain activity during observation of actions. Influence of action content and subject's strategy. Brain 120: 101–115

De Renzi E (1989) Apraxia. In: Boller F, Grafman J (eds) Handbood of neuropsychology. Vol 2. Elsevier Science Publishers, Amsterdam, pp 245–263

Di Pellegrino G, Fadiga L, Fogassi L, Gallese V, Rizzolatti G (1992) Understanding motor events: a neurophysiological study. Exp Brain Res 91: 176–180

Dittrich WH (1993) Action categories and the perception of biological motion. Perception 22: 15–22

Duhamel JR, Colby CL, Goldberg ME (1992) The updating of the representation of visual space in parietal cortex by intended eye movements. Science 255: 90–92

Fadiga L, Fogassi L, Pavesi G, Rizzolatti G (1995) Motor facilitation during action observation: a magnetic stimulation study. J Neurophysiol 73: 2608–2611

Faillenot I, Toni I, Decety J, Jeannerod M (1997) Visual pathways for object-oriented action and object identification. Functional anatomy with PET. Cereb Cortex 7: 77–85

Feltz DL, Landers DM (1983) The effects of mental practice on motor skill learning and performance. A meta-analysis. J Sport Psychol 5: 25–57

Fletcher PC, Happe F, Frith U, Baker SC, Dolan RJ, Frackowiak RSJ, Frith CD (1995) Other minds in the brain: a functional imaging study of "theory of mind" in story comprehension. Cognition 57: 109–128

Fodor JA (1983) The modularity of mind. MIT Press, Cambridge, MA

Frith U (1989) Autism: explaining the enigma. Blackwell, Oxford

Gallese V, Fadiga L, Fogassi, Rizzolatti G (1996) Action recognition in the premotor cortex. Brain 119: 593–609

Goel V, Grafman J, Sadato N, Hallett M (1995) Modeling other minds. NeuroReport 6, 1741–1746

Goodale MA (1993) Visual pathways supporting perception and action in the primate cerebral cortex. Curr Opin Neurobiol 3: 578–585

Goodale MA (1997) Visual routes to perception and action in the cerebral cortex. In: Boller F, Grafman J (eds) Handbook of Neuropsychology. Vol 11. Action and Cognition. Elsevier, Amsterdam, pp 91–110

Goodale MA, Milner AD, Jakobson LS, Carey DP (1991) A neurological dissociation between perceiving objects and grasping them. Nature 349: 604–610

Grafman J (1989) Plans, actions and mental sets. Managerial knowledge units in the frontal lobes. In: Perecman E (ed) Integrative theory and practice in clinical neuropsychology. Erlbaum, Hillsdale, pp 93–138

Grèzes J, Costes N, Decety J (1998) Top down effect of strategy on the perception of human biological motion: A PET investigation. Cognitive Neuropsychology 15: in press

Halligan PW, Marshall JC, Wade DT, Davey J, Morrisson D (1993) Thumb in cheek? Sensory reorganization and perceptual plasticity after limb amputation. NeuroReport 4: 233–236

Happé F, Ehlers S, Fletcher P, Frith U, Johannsson M, Gillberg C, Dolan R, Frackowiak R, Frith C (1996) Theory of mind in the brain. Evidence from PET scan study of Asperger syndrome. NeuroReport 8: 197–201

Haxby JV, Grady CL, Horwitz B, Ungerleider LG, Mishkin RE, Carson RE, Herscovitch P, Shapiro MB, Rapoport SI (1991) Dissociation of object and spatial visual processing pathways in human extrastriate cortex. Proc Nat Acad Sci USA 88: 1621–1625

Heilman KM, Watson RT, Rothi LG (1997) Disorders of skilled movements: limb apraxia. In: Feinberg TE, Farah MJ (eds) Behavioral neurology and neuropsychology. McGraw-Hill, New York, pp 227–235

Jeannerod M, Decety J, Michel F (1994) Impairment of grasping movements following bilateral posterior parietal lesion. Neuropsychologia 32: 369–380

Jeannerod M, Decety J (1995) Mental motor imagery: a window into the representational stages of action. Curr Opin Neurobiol 5: 727–732

Johansson G (1973) Visual perception of biological motion and a model for its analysis. Perception Psychophysics 14: 201–211

Kalaska JF, Crammond DJ (1995) Deciding not to go: neural correlates of response selection in a go/nogo task in primate premotor and parietal cortex. Cereb Cortex 5: 410–428

Köhler S, Kapur S, Moscovitch M, Winocur G, Houle S (1995) Dissociation of pathways for object and spatial vision: a PET study in humans. Neuroreport 6: 1865–1868

Konorski J (1967) Integrative activity of the brain: an interdisciplinary approach. University of Chicago, Chicago

Kosslyn SM (1997) Mental imagery. In: Gazzaniga MS (ed) Conversations in the cognitive neurosciences. MIT Press, Cambridge, pp 155–174

Kozlowski LT, Cutting JE (1977) Recognizing the sex of a walker from point-lights display. Perception Psychophysics 21: 575–580

Lang W, Petit L, Höllinger P, Pietrzyk U, Tzourio N, Mazoyer B, Berthoz A (1994) A positron emission tomography study of oculomotor imagery. NeuroReport 5: 921–924

Leonardo M, Fieldman J, Sadato N, Campbell G, Ibanez V, Cohen L, Deiber MP, Jezzard P, Pons T, Le Bihan D, Hallett M (1995) A functional magnetic resonance imaging study of cortical regions associated with motor task execution and motor ideation in humans. Human Brain Mapping 3: 83–92

Leslie AM (1987) Pretense and representation: the origins of "theory of mind". Psychol Rev 94: 412–426

Meltzoff AN, Moore MK (1992) Early imitation within a functional framework: the importance of person identity, movement, and development. Infant Behav Develop 15: 479–505

Meltzoff AN, Gopnik A (1993) The role of imitation in understanding persons and developping a theory of mind. In: Baron-Cohen S, Tager-Flusberg H, Cohen DJ (eds) Understanding other minds. Oxford Medical Publications, New York, pp 335–366

Miller EK, Erickson CA, Desimone R (1996) Neural mechanisms of visual working memory in prefrontal cortex of the macaque. J Neurosci 16: 5154–5167

Milner AD, Goodale MA (1995) The visual brain in action. Oxford University Press, Oxford

Mogilner A, Grossman JAI, Ribary U, Joliot M, Volkmann J, Rapaport D, Beasley RW, Llinas RR (1993) Somatosensory cortical plasticity in adult humans revealed by magnetoencephalography. Proc Nat Acad Sci USA 90: 3593–3597

Orliaguet JP, Kandel S, Boë LJ (1998) Visual perception of motor anticipation in cursive handwritting: influence of spatial and movement information on the prediction of forthcoming letters. Perception, in press

Parsons LM, Fox PT, Downs JH, Glass T, Hirsch TB, Martin CC, Jerabek PA, Lancaster JL (1995) Use of implicit motor imagery for visual shape discrimination as revealed by PET. Nature 375: 54–58

Pascual-Leone A, Dang N, Cohen LG, Brasil-Neto JP, Cammarota A, Hallett M (1995) Modulation of muscle responses evoked by transcranial magnetic stimulation during the acquisition of new fine motor skills. J Neurophysiol 74: 1037–1045

Passingham RE (1996) Functional specialization of the supplementary motor area in monkeys and humans. In: Lüders HO (ed) Advances in Neurology. Vol 70. Supplementary sensorimotor area. Lippincott-Raven, Philadelphia, pp 105–116

Perrett DI, Mistlin AJ, Chitty AJ (1987) Visual cells responsive to faces. Trends Neurosci 10: 358–364

Perrett DI, Harries MH, Bevan R, Thomas S, Benson PJ, Mistlin AJ, Chitty AJ, Hietanen JK, Ortega JE (1989) Frameworks of analysis for the neural representation of animate objects and actions. J Exper Biol 146: 87–114

Perrett DI, Mistlin AJ, Harries MH, Chitty AJ (1990) Understanding the visual appearance and consequence of actions. In: Goodale MA (ed) Vision and action. Ablex Publishing Corporation, Norwood, New Jersey, pp 163–180

Perrett DI, Emery NJ (1994) Understanding the intentions of others from visual signals: neurophysiological evidence. Cahiers Psychol Cogn 13: 683–694

Premack D, Woodruff G (1978) Does the chimpanzee have a theory of mind? Behav Brain Sci 1: 515–526

Preuss TM (1995) The argument from animals to humans in cognitive neuroscience. In: Gazzaniga MS (ed) The cognitive neurosciences. MIT Press, Cambridge, pp 1227–1241

Ramachandran VS, Stewart M, Rogers-Ramachandran DC (1992) Perceptual correlates of massive cortical reorganization. NeuroReport 3: 583–586

Rao SC, Rainer G, Miller EK (1997) Integration of what and where in the primate prefrontal cortex. Science 276: 821–824

Rizzolatti G, Fadiga L, Gallese V, Fogassi L (1996a) Premotor cortex and the recognition of motor actions. Cogn Brain Res 3: 131–141

Rizzolatti G, Fadiga L, Matelli M, Bettinardi V, Paulesu E, Perani D, Fazio F (1996b) Localization of grasp representations in humans by PET: 1. Observation versus execution. Exper Brain Res 111: 246–252

Roland PE, Skinhoj E, Lassen NA, Larsen B (1980) Different cortical areas in man in organization of voluntary movements in extrapersonal space. J Neurophysiol 43: 137–150

Roth M, Decety J, Raybaudi M, Massarelli R, Delon-Martin C, Segebarth C, Decorps M, Jeannerod M (1996) Possible involvement of primary motor cortex in mentally simulated movement: A functional magnetic resonance imaging study. NeuroReport 7: 1280–1284

Rothi LJG, Heilman KM (1997) Apraxia: the neuropsychology of action. Psychology Press, Hove UK

Rothi LJG, Mack L, Heilman KM (1986) Pantomime agnosia. J Neurol Neurosurg Psychiat 49: 451–454

Rothi LJG, Ochipa C, Heilman KM (1991) A cognitive model of limb praxis. Cogn Neuropsychol 8: 443–458

Runeson S, Frykholm G (1983) Kinematic specifications of dynamics as an informational basis for person-and-action perception: Expectation, gender recognition, and deceptive intention. J Exper Psychol General 112: 585–615

Sakata H, Taira M, Kusunoki M, Murata A, Tanaka Y (1997) The parietal association cortex in depth perception and visual control of hand action. Trends Neurosci 20: 350–357

Shiffrar M, Freyd JJ (1993) Timing and apparent motion path choice with human body photographs. Psychol Sci 4: 379–384

Sirigu A, Cohen L, Duhamel JR, Pillon B, Dubois B, Agid Y, Pierrot-Deseilligny C (1995) Congruent unilateral impairments for real and imagined hand movements. Neuroreport 6: 997–1001

Sirigu A, Duhamel JR, Cohen L, Pillon B, Dubois B, Agid Y (1996) The mental representation of hand movements after parietal cortex damage. Science 273: 1564–1568

Smith IM, Bryson SE (1994) Imitation and action in autism. Psychol Bull 116: 259–273

Stepahn KM, Fink GR, Passingham RE, Silbersweig D, Ceballos-Baumann AO, Frith CD, Frackowiak RSJ (1995) Functional anatomy of the mental representation of upper extremity movements in healthy subjects. J Neurophysiol 73: 373–386

Sumi S (1984) Upside-down presentation of the Johansson moving light-spot pattern. Perception 13: 283–286

Tooby J, Cosmides L (1995) Mapping the evolved functional organization of mind and brain. In: Gazzaniga M (ed) The cognitive neurosciences. MIT Press, Cambridge MA, pp 1185–1197

Ungerleider LG (1995) Functional brain imaging studies of cortical mechanism for memory. Science 270: 769–775

Ungerleider LG, Mishkin M (1982) Two visual systems. In: Ingle DJ, Goodale MA, Mansfield RJW (eds) Analysis of visual behavior. MIT Press, Cambridge MA, 549–586

Verfaillie K (1993) Orientation-dependent priming effects in the perception of biological motion. J Exper Psychol Human Perception Performance 19: 992–1013

Viviani P, Stucchi N (1989) The effect of movement velocity on form perception: geometric illusions in dynamic display. Perception Psychophysics 46: 266–274

Vogt S (1996) Imagery and perception-action mediation in imitative actions. Cogn Brain Res 3: 79–86

Hofsten C von (1993) Prospective control: a basic aspect of action development. Human Devel 36: 253–270

Watson JDG, Myers R, Frackowiak RSJ, Hajnal JV, Woods RP, Mazziotta JC, Shipp S, Zeki S (1993) Area V5 of the human brain: evidence from a combined study using PET and MRI. Cereb Cortex 3: 79–94

Wellman HM (1993) Early understanding of mind: the normal case. In: Baron-Cohen S, Tager-Flusberg H, Cohen DJ (eds) Understanding other minds. Oxford Medical Publications, New York, pp 10–39

Xerri C, Stern JM, Merzenich MM (1994) Alterations of the cortical representation of the rat ventrum induced by nursing behavior. J Neurosci 14: 1710–1721

Yue G, Cole KJ (1992) Strength increases from the motor program: comparison of training with maximal voluntary and imagined muscle contractions: J Neurophysiol 67: 1114–1123

Evidence for Four Forms of Neuroplasticity

J. Grafman, Ph. D.[1] and *I. Litvan, M. D.*[2]

Summary

We suggest that at least four major forms of functional neuroplasticity can be studied in normal human subjects and patients. The four forms of functional neuroplasticity are homologous area adaptation, cross-modal reassignment, map expansion, and compensatory masquerade. Homologous area adaptation is the assumption of a particular cognitive process by a homologous region in the opposite hemisphere. Cross-modal reassignment occurs when structures previously devoted to processing a particular kind of sensory input now accept input from a new sensory modality. Map expansion is the enlargement of a functional brain region on the basis of performance. Compensatory masquerade is a novel allocation of a particular cognitive process to perform a task. By focusing on these four forms of functional neuroplasticity, several fundamental questions about how functional cooperation between brain regions is achieved can be addressed.

Introduction

Recent research in cognitive neuroscience has made great strides in mapping the functions of the human brain and in determining the knowledge, representational elements, and processes that are subserved by cognitive maps (Merzenich et al. 1996a). Although much work remains to be done to provide a mature functional cartographic atlas of the cerebral cortex and subcortical structures, some assumptions about how representational knowledge is stored are generally agreed upon. The brain appears to be composed of modular neural networks (for the purposes of this chapter, their "penetrability" will be ignored) within which a defined representational unit is homogeneously represented. Such units may range from edge detectors, used in visual processing (Gilbert 1996) and stored in the occipital cortex, to high-level plans used to guide behavior and stored in

[1] Chief, Cognitive Neuroscience Section, Medical Neurology Branch, National Institute of Neurological Disorders and Stroke, National Institutes of Health, Bethesda, Maryland USA
[2] Chief, Neuropharmacology Unit, Defense and Veterans Head Injury Program, Henry M. Jackson Foundation, Rockville, Maryland USA

J. Grafman / Y. Christen (Eds.)
Neuronal Plasticity:
Building a Bridge from the Laboratory to the Clinic
© Springer-Verlag Berlin Heidelberg New York 1999

Function of Region B Transferred to Region A   **Region A Accepts New Modality Input**

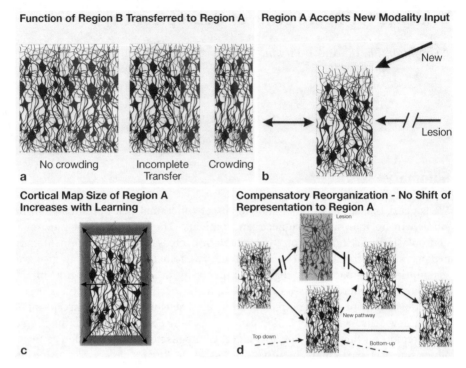

Fig. 1. Illustration of four major forms of neuronal system plasticity. The first form of plasticity (**a**) indicates that, during childhood, the functions of a particular region may be literally transferred to another brain area (homologous region adaptation). The second form (**b**) indicates that a particular brain region may accept input from another modality that, although ordinarily processed elsewhere in the brain, is now diverted to that region for specific information processing purposes (cross-modal reassignment). The third form of plasticity (**c**) indicates that changes in cortical map topography can occur for a variety of reasons, including skill learning and passive invasion of neighboring cortical tissue following trauma to that tissue or its input system (map expansion). The fourth form of plasticity (**d**) indicates that a spared brain region has assumed a primary role in performing a task after an injury to a focal brain region that previously played the prominent processing role (compensatory masquerade). See the text for more details.

the prefrontal cortex (Grafman 1995). To perform a typical human function, sets of modules are cooperatively activated. Thus, the human brain can be subdivided into its constituent functional (subcomponent) modules, sets of which must be combined in order to perform a task.

Current thinking about neuroplasticity suggests that there are at least *four major kinds of potential neuroplastic changes* operating at the representational module level (Fig. 1a–d). These are *homologous area adaptation, cross-modal reassignment, map expansion,* and *compensatory masquerade.*

Four Forms of Neuroplasticity

One form of neuroplasticity, *homologous area adaptation,* appears most active during an early, critical stage of human development and underlies the notion that damage to a particular brain region and its cognitive operation(s) can be compensated for by shifting the individual (or set of) operation(s) to other brain areas that do not include the affected module (e.g., the function is usually shifted to another module in the homologous region of the opposite hemisphere; Chugani et al. 1996). The result of this neuroplastic change has led to the idea that the new brain area accepting a new cognitive operation is now more "crowded" with distinct cognitive representations. This crowding leads to a sparser representation of knowledge within the shifted module and increases the likelihood of dual-task interference when two tasks to be performed simultaneously involve adjacent "modules" in the cortex – one of which has been shifted from its former and "natural" location in the brain. This form of neuroplastic change is reported less often in adults.

In a single-case study (Levin et al. 1996), we studied an adolescent who had incurred a severe right parietal lobe brain injury as a young child. Despite the severity and location of the injury, our evaluation showed that the adolescent had developed relatively normal visuospatial skills but had impaired arithmetic skills. The inference is that, at the time of the injury, the left parietal region assumed some of the responsibilities of the functions normally stored in the right parietal lobe. Since much of arithmetic computation is learned in school, the injury and plasticity occurred prior to the age of arithmetic acquisition. Thus, we argued that spatial processes had claimed the left parietal region prior to arithmetic instruction, making it much more difficult for the patient to learn and store arithmetic facts. In essence, there was little room left in the left parietal lobe for the storage of arithmetic facts and concepts. Functional magnetic resonance imaging of this patient during arithmetic processing indicated that, in fact, he activated left parietal tissue (among other regions), showing that that region was still genetically programmed to store arithmetic facts even if it was now more committed to spatial processing.

Some investigators have claimed that there is a relationship between the proportion of a functional region that is damaged and the amount of homologous region adaptation that can occur. The homologous region to the one that is damaged is only able to reorganize and assume a new function when contralateral inhibitory input is removed. Given this logic, it would be more useful to have complete rather than incomplete damage to a region where a primary function was represented in order for an optimal transfer of function to a homologous area to occur.

We recently had a chance to study another patient who suffered a severe stroke destroying almost the entire left hemisphere, although he had some spared, functionally active islands in the left parietal and frontal cortices (Basso et al., manuscript in preparation; Grafman et al., manuscript in preparation). This patient could read words but not nonwords. He also had great difficulty performing calculations. We studied him while he was performing behavioral tasks with functional magnetic resonance imaging. Word reading activated a broadly distributed network in the right hemisphere. Attempts to read nonwords activated widely scattered and punctate areas in the left hemisphere. This finding suggests that the right hemisphere could assume some functions of the left hemisphere after massive damage. It seems clear that the major reading pathways in the left hemisphere were destroyed, allowing the right hemisphere to assume its reading functions; however, phonological construction required for nonwords could not be transferred, highlighting some of the limitations of this form of functional neuroplasticity in adulthood. Interestingly, during fMRI scanning while the patient was performing arithmetic fact verification, we found both right and left parietal lobe activation. Even with this bilateral activation pattern, the patient performed poorly, although his performance slightly increased in accuracy after a week's worth of training (with a concomitant increase in cortical activation). It can be inferred that, in this case, the remaining left parietal tissue was inhibiting the right parietal cortex from assuming more of a role in calculation, leading to a less-than-adequate (if at all) transfer of function to the right hemisphere. Although this case study is suggestive of the limits of homologous region adaptation, much more work is needed to confirm this speculative hypothesis, since an alternative view could be that both areas are needed to perform this complex task.

A second form of neuroplasticity, *cross-modal reassignment*, involves the introduction of new inputs into a representational brain region that has been deprived of its main inputs. For example, PET and fMRI studies of tactile discrimination ability have shown that subjects who became blind early in childhood, but were tested as adults, have somatosensory input redirected into area V1 of the occipital cortex, whereas normal controls do not show evidence of any V1 activation during the same task (Sadato et al. 1996). Perhaps, in the blind, such input succeeds in activating the representations stored in area V1 because such representations are in an "abstract format", that is, the cognitive operations are independent of the modality of input. For example, discrimination of meaningful geometric forms (such as Braille letters) could occur in previously defined primary visual areas if the new modality of input required the same kind of geometric form discrimination that is ordinarily handled by the "visual system." In the study referred to, only tactile discrimination of raised Braille dots that revealed geometric form (as opposed to simply passing the hand across a raised homogeneous field of tactile stimuli) or language activated V1 in the blind. In contrast, in normal subjects exposed to the same stimuli and tasks, not only was V1 activation not found, but there was evidence of decreased activation in area V1, suggesting that attention devoted to brain regions processing the tactile features of stimuli (e.g., parietal cortex) ordinarily inhibit competitive brain systems (such as the visual system). This inhibitory activity also suggests that – even in adults – there may be a pre-existing pathway (currently used for inhibition) that potentially could be transformed for facilitation during the processing of alternatively presented information (such as tactual information).

A recent study using transcranial magnetic stimulation over the occipital cortex in blind subjects (Cohen et al. 1997) was also able to demonstrate interference with Braille reading in subjects blind from birth or shortly afterwards (within a few years of birth). However, there are probably limitations to the feasibility of this form of neuroplasticity. For example, color processing cells in the occipital cortex are specialized for visual input and would be unlikely to accept other forms of input.

The third form of neuroplasticity, *map expansion*, demonstrates the flexibility of brain regions devoted to a particular kind of knowledge or cognitive operation. Recent work has indicated that the size of cortical maps devoted to a particular information processing function may enlarge with skilled practice or frequent exposure to a stimulus. In two studies using two different techniques (transcranial magnetic stimulation and electroencephalography; Pascual-Leone et al. 1994; Zhuang et al. 1997), we demonstrated that implicit learning of a visuomotor sequence induced sensorimotor map expansion in the early stages of (implicit) learning. When learning became explicit, the cortical map size returned to baseline. This enlargement can be quickly seen over the first few minutes of practice or exposure. There is also evidence that this rapid enlargement of selected cortical maps can be persistent in individuals who develop or are trained in a particular skill that they need to utilize on a routine basis (Rosenzweig and Bennett 1996). The meaning of map expansion is still unclear (Donoghue 1995). It is possible that it could have two implications. One is that, with use, the cortical region devoted to a particular operation can expand into other regions usually dedicated to another function, in essence recruiting new neurons into the network. However, over practice, sometimes regions show increased activation and sometimes regions show decreasing activation. Another implication of map expansion follows from these observations. That is, when the exact unit of representation to be used to process the bottom-up or top-down information is still undecided, the entire network needs to be active. When the exact unit of representation is selected, the network can relax and less energy is expended.

The fourth form of neuroplasticity, *compensatory masquerade*, means that the novel use of an established, but intact, cognitive process to perform a task previously dependent upon an impaired cognitive process has occurred. This can be an insidious process whose discovery is dependent upon the utilization of fine-grained cognitive tasks. For example, there may be many ways to navigate a route from home to the office. One way may depend most upon a sense of spatial coordinates, is relatively implicit, and is performed rapidly. Another way may depend on verbal labeling of landmarks, is relatively explicit, and is performed more slowly. A brain injury may affect either process but spare one. The patient may then be able, over a short period of time, to use the strategy that is spared to navigate the same path. Unless a neuropsychological study evaluated both processes in some detail, the investigator could be misled into thinking that a more basic form of neuroplasticity had occurred (e.g., homologous area adaptation) rather than a realigned distributed network composed of previously existing modules that played a previously minor role in performing the task now catapulted into a major role.

Discussion

These four major forms of neuroplasticity can be studied in normal human subjects and patients. Other levels of neuroplasticity that have been, and continue to be, thoroughly studied in in vivo and in vitro preparations include structural changes at the synaptic junction following repeated stimulation (Buonomano et al. 1997; Debanne 1996; Frotscher et al. 1997; Gu 1995; Ide et al. 1996; Kirkwood and Bear 1995). However, we believe that by focusing on the four forms of plasticity outlined above, several fundamental questions about how functional cooperation between brain regions is achieved can be asked. The implication of answering these questions for learning potential and recovery of function from brain injuries is profound, as we outline below.

Why does shifting the functional operations of a particular brain region to a more distant site seem to decline over development? We hypothesize that this may be directly due to increasing connectivity between maturing cortical and subcortical areas dedicated to a particular cognitive or information processing operation. The increasing magnitude and stability of the connectivity implies that highly sensitive distributed networks are formed which, when co-activated, lead to the performance of varied cognitive tasks. If functional substitution was permitted beyond the early stages of the formation of these distributed but connected and functionally committed modules, then a rather chaotic situation would emerge that could potentially require continual (and substantial) reorganization of distributed networks (potentially resembling the somewhat random and incompetent wiring of telephone lines that littered downtown Buenos Aires in the last decade). This situation would prove to be extraordinarily ineffective and disruptive to the ongoing process of learning and memory that development entails. Thus, we predict a direct association between the solidifying of connections between components of a distributed network and a decline in functional substitution ability of the involved components.

A second hypothesis about plasticity emerges from the similarity in certain cortical operations across apparently modality-specific regions of sensory cortex, as described above as a second form of neuroplasticity. In the example we used, the occipital lobe, comprising several cortical areas concerned with discriminating among vertical and horizontal features of objects, was deprived of visual input. In this circumstance, should one expect atrophy and lack of activation across V1? There is now evidence that tactile input could be processed in area V1 if discrimination of the same stimulus properties was required, i.e., discrimination of width, angles, and feature conjunctions. In this case, a stable module (in terms of retaining its basic connecting paths to and from other cortical modules) takes over the processing of geometric stimuli, even though the input is now tactile and not visual. Thus, no dramatic restructuring of distributed networks is necessary as further, more complex processing of such stimuli proceeds normally through the network.

The third form of neural plasticity reflects a functional "balkanization" of the cerebral cortex regions devoted to various cognitive operations. That is, there is a natural competitiveness among cortical regions which is reflected in constantly shifting boundaries between local cortical areas. This boundary shift is rather non-disruptive during normal learning, although an area that subserves a skill that is particularly valued by the person (e.g., guitar playing) can, over time, capture more cortical territory (e.g., cortical maps subserving finger representation in the hand that forms chords or plucks strings may enlarge). In the case of peripheral damage to a limb, cortical regions adjacent to territory representing the now lost limb invade the lost limb's previous cortical territory, staking out claims for the now uncommitted neurons (Aglioti et al. 1994; Melzack et al. 1997). Naturally, the most functionally active of the adjacent cortical regions will capture more cortical territory. However, some neurons may be activated by top-down processing, driving them to reincarnate, through mental imagery, images of the limb and sensation of its previous motoric capacity. Finally, when partial loss of an area occurs as a result of brain damage, adjacent cortical regions also may overlap with the damaged regions, affecting the ability of the patient to recover and perhaps causing interference during the simultaneous performance of tasks, requiring the participation of both affected and unaffected adjacent cortical regions.

In summary, we argue that there is evidence for four forms of functional plasticity: the first, homologous region adaptation, which is limited by the development of mature distributed neural networks; the second form allows novel sensory input into a module or set of modules previously receiving evolutionary determined primary sensory input; the third is present throughout the lifespan and allows the normal ebb and flow of modular boundaries depending on the frequency of use of that module in cognitive, motor, or sensory processing; and the fourth, compensatory masquerade, which is simply a reorganization of pre-existing functional neuronal networks.

There are many contemporary frameworks for conceptualizing neuroplasticity (DeLisi 1997; Goldman and Plum 1997; Kapur 1996; Leviton et al. 1995; Luders et al. 1997; Mirmiran et al. 1996; Plaut 1996; Seil 1997; Seitz and Freund 1997; Wolf 1996). Given the view of functional plasticity we subscribe to, in what direction should future research proceed? We suggest that emphasizing map topography manipulation and rerouting of input systems would give cognitive neuroscience research efforts the biggest immediate payoff. It is clear that map changes occur with learning and use, but the limitations of such topographic changes in functional reorganization are unknown. Are the map changes governed by a neurogeometric principle that is determined by the distance of neuronal columns from the epicenter of the neighboring functional region? Perhaps that geometric variable interacts with the overall size of the functional region as well as with the residual activation of neighboring functional regions to determine its neuroplastic capability. In any case, these hypotheses can be tested in both normal subjects (during learning and dual-task paradigms that activate neighboring functional maps) as well as patients (looking at training-induced versus passively induced changes in cortical map size) and should be relevant for measuring functional improvement during and after participation in cognitive rehabilitation programs (Irvine and Rajan 1996; Merzenich et al. 1996b). The input substitution approach is also potentially very promising. It may be able to tell us which functional areas are able to accept a variety of primary sensory inputs based on a common input signal or that the receiving brain region is less concerned with the input modality than with processing a certain kind of coded input (e.g., geometric form). Although such a region may have a primary input based on evolutionary development, it may be able to accept atypical sensory input because of the similarity of the information being processed. These cross-modal reassignment studies can be administered to normal subjects (using sensory or cognitive deprivation paradigms) and patients with acquired sensory lesions (e.g., blind or deaf subjects).

Obtaining experimental control over paradigms designed to manipulate functional neuroplasticity in normal subjects and patients is advantageous since it allows for control over, and specification of, the conditions under which functional neuroplasticity can occur (Rauschecker 1997). This kind of control also allows for the eventual development of "challenge" studies using pharmacologic agents to facilitate or inhibit any functional neuroplastic changes that can be identified using the experimental procedures suggested in the previous paragraph. Indeed, the future looks promising as we probe the limits of functional neuroplasticity in the adult human while answering fundamental questions about the stable and dynamic topography of cortical and subcortical information processing maps.

References

Aglioti S, Bonazzi A, Cortese F (1994) Phantom lower limb as a perceptual marker of neural plasticity in the mature human brain. Proc R Soc Lond B Biol Sci 255:273–278

Buonomano DV, Hickmott PW, Merzenich MM (1997) Context-sensitive synaptic plasticity and temporal-to-spatial transformations in hippocampal slices. Proc. Natl Acad Sci USA 94:10403–10408

Chugani HT, Muller RA, Chugani DC (1996) Functional brain reorganization in children. Brain Dev 18:347–356

Cohen LG, Celnik P, Pascual-Leone A, Corwell B, Falz L, Dambrosia J, Honda M, Sadato N, Gerloff C, Catala MD, Hallett M (1997) Functional relevance of cross-modal plasticity in blind humans. Nature 389:180–183

Debanne D (1996) Associative synaptic plasticity in hippocampus and visual cortex: cellular mechanisms and functional implications. Rev Neurosci 7:29–46

DeLisi LE (1997) Is schizophrenia a lifetime disorder of brain plasticity, growth and aging? Schizophr Res 23:119–129

Donoghue JP (1995) Plasticity of adult sensorimotor representations. Curr Opin Neurobiol 5:749–754

Frotscher M, Heimrich B, Deller T (1997) Sprouting in the hippocampus is layer-specific. Trends Neurosci 20:218–223

Gilbert CD (1996) Plasticity in visual perception and physiology. Curr Opin Neurobiol 6:269–274

Goldman S, Plum F (1997) Compensatory regeneration of the damaged adult human brain: neuroplasticity in a clinical perspective. Adv Neurol 73:99–107

Grafman J (1995) Similarities and distinctions among current models of prefrontal cortical functions. Ann N Y Acad Sci 769:337–368

Gu Q (1995) Involvement of nerve growth factor in visual cortex plasticity. Rev. Neurosci 6:329–351

Ide CF, Scripter JL, Coltman BW, Dotson RS, Snyder DC, Jelaso A (1996) Cellular and molecular correlates to plasticity during recovery from injury in the developing mammalian brain. Prog Brain Res 108:365–377

Irvine DR, Rajan R (1996) Injury- and use-related plasticity in the primary sensory cortex of adult mammals: possible relationship to perceptual learning. Clin Exp Pharmacol Physiol 23:939–947

Kapur N (1996) Paradoxical functional facilitation in brain-behaviour research. A critical review. Brain 119 (Pt 5):1775–1790

Kirkwood A, Bear MF (1995) Elementary forms of synaptic plasticity in the visual cortex. Biol Res 28:73–80

Levin HS, Scheller J, Rickard T, Grafman J, Martinkowski K, Winslow M, Mirvis S (1996) Dyscalculia and dyslexia after right hemisphere injury in infancy. Arch Neurol 53:88–96

Leviton A, Bellinger D, Pagano M, Rabinowitz M (1995) Models of delayed recovery. J Child Neurol 10:385–391

Luders HO, Comair YG, Bleasel AF, Holthausen H (1997) Recovery of function following lesions of eloquent brain areas. Adv Neurol 73:335–346

Melzack R, Israel R, Lacroix R, Schultz G (1997) Phantom limbs in people with congenital limb deficiency or amputation in early childhood. Brain 120 (Pt 9):1603–1620

Merzenich M, Wright B, Jenkins W, Xerri C, Byl N, Miller S, Tallal P, Miller SL, Bedi G, Byma G, Wang X, Nagarajan SS, Schreiner C. Jenkins WM, Merzenich MM (1996a) Cortical plasticity underlying perceptual, motor, and cognitive skill development: implications for neurorehabilitation. Language comprehension in language-learning impaired children improved with acoustically modified speech [see comments]. Cold Spring Harb Symp Quant Biol 61:1–8

Merzenich MM; Jenkins WM, Johnston P, Schreiner C, Miller SL, Tallal P (1996b). Temporal processing deficits of language-learning impaired children ameliorated by training [see comments]. Science 271:77–81

Mirmiran M, van Someren EJ, Swaab DF (1996) Is brain plasticity preserved during aging and in Alzheimer's disease? Behav Brain Res 78:43–48

Pascual-Leone A, Grafman J, Hallett M (1994) Modulation of cortical motor output maps during development of implicit and explicit knowledge [see comments]. Science 263:1287–1289

Plaut DC (1996) Relearning after damage in connectionist networks: toward a theory of rehabilitation. Brain Lang 52:25–82

Rauschecker JP (1997) Mechanisms of compensatory plasticity in the cerebral cortex. Adv Neurol 73:137–146

Rosenzweig MR, Bennett EL (1996). Psychobiology of plasticity: effects of training and experience on brain and behavior. Behav Brain Res 78:57–65

Sadato N, Pascual-Leone A, Grafman J, Ibanez V, Deiber MP, Dold G, Hallett M (1996) Activation of the primary visual cortex by Braille reading in blind subjects [see comments]. Nature 380:526–528

Seil FJ (1997) Recovery and repair issues after stroke from the scientific perspective. Curr Opin Neurol 10:49–51

Seitz RJ, Freund HJ (1997) Plasticity of the human motor cortex. Adv Neurol 73:321–333

Wolf S (1996) Using our brains [editorial]. Integr Physiol Behav Sci 31:195–201

Zhuang P, Toro C, Grafman J, Manganotti P, Leocani L, Hallett M (1997) Event-related desyncronization (ERD) in the alpha frequency during development of implicit and explicit learning. Electroencephal Clin Neurophysiol 102:374–381

Imaging Investigations of Human Brain Plasticity

R. S. J. Frackowiak*

In the last quarter of this century major advances in computing and mathematics have led to remarkable opportunities for new studies of the structure and function of the human brain.

The Present State of Neuroimaging Methods

The Aims of Non-Invasive Human Brain Mapping

An aim of the new discipline of imaging neuroscience is to describe the functional organization of the human brain at the level of large neuronal groupings, networks and systems and to relate these to human behaviour, perception, action and consciousness. A systems level of description addresses how integrated brain functions are embodied in the physical structure of the brain (Frackowiak et al. 1997).

Sensory input and motor output are arranged in topographically organised maps in the brain. A description of the location and interaction of such maps in normal and damaged human brain would represent a significant increase in knowledge to add to that drawn from psychology and neuropsychological observations in patients with brain damage. Functional homologies between non-human primate brain and human brain are controversial but critically relevant to how far rules governing functional brain organization determined in one species can be transposed to another. For example, the massive development of frontal and parietal lobes and evolution of language are unique to human brains, and yet we know comparatively little in detail of the functional attributes of the frontal and parietal lobes in man.

Brain organisation can be understood at various levels in both temporal and spatial domains. Neurons, neuronal groupings, modules and larger scale, functionally homogeneous or cooperative neuronal populations represent different spatial levels of brain organization. Normal brains exhibit electrical activity with a periodicity or miliseconds, hundreds of milliseconds or even seconds. For example, the action potential, evoked potential, readiness potential and delta wave of electroencephalography are all examples of different temporal levels of

* Wellcome Department of Cognitive Neurology, Institute of Neurology, Green Square, London, England UK

J. Grafman / Y. Christen (Eds.)
Neuronal Plasticity:
Building a Bridge from the Laboratory to the Clinic
© Springer-Verlag Berlin Heidelberg New York 1999

organization. Modern non-invasive brain imaging techniques bring a number of these levels of organization within the range of measurement and so provide a means for systematic analysis of the spatio-temporal functional architecture of the human brain and also of plasticity of that architecture.

What is the Present State of Functional Mapping Technologies?

Most new neuroscientific information has been generated by PET imaging because functional magnetic resonance imaging (fMRI) is still in active development. There are a number of outstanding issues that will, for the present, determine which modality is more appropriate for answering a given scientific question. PET measures the distribution of brain activity simultaneously in all regions. fMRI collects images slice by slice with a sequential, though potentially extremely rapid, scan of the image space (e.g., 60 msec per slice, with 60 3-mm thick slices to cover the whole brain, giving a total scanning time of 3.6 secs). The effects of non-simultaneity and randomisation of stimulus onset to the first scanned plane on analysis of temporal characteristics of brain activity are being solved. It is also now possible to average repeated event-related scans as in classical neuro-physiology. There is therefore more dynamic potential with fMRI than with PET.

Absolute and relative increases and decreases of perfusion can be monitored with radio-tracer based perfusion methods such as PET. It is thought that the majority of blood oxygen level dependent (BOLD) signal in fMRI is due to an uncoupling of oxygen metabolism from oxygen delivery to activated brain tissue. The resultant local increase of oxyhemoglobin, relative to deoxyhemoglobin in activated brain tissue, produces an image signal and relative changes in this signal can be recorded and interpreted. Newer fMRI methods are being developed to give a purer perfusion signal than that obtainable with BOLD. Whether these methods become useful depends on whether the whole brain can be imaged and on the comparative sensitivity of the various methods.

The precise localisation of sites of activation is fundamental to accurate brain mapping. There is controversy whether the majority of fMRI signal arises in locally activated tissue or from surrounding draining veins. It is possible to optimize fMRI acquisition parameters to record signal predominantly from small vessels, thus reflecting flow rather than blood volume-induced changes. One such technique (asymmetric spin-echo scanning) results in signal that has a reliable perfusion dependency, but there is some loss of sensitivity that may render the detection of small signal changes from activated brain tissue difficult. Another possibility is that angiographic MRI images, which record the position of cerebral blood vessels, can be used to correct functional (gradient-echo) images for signal artifacts generated by saturated blood in veins draining an activated brain area. An associated issue, that of the generation of image artifacts by movement of the brain during and between scans in the MRI camera's field-of-view, has been largely overcome by fast echo-planar imaging (EPI) and *post hoc* image re-

registration software. It is, however, very difficult to correct for stimulus-correlated movement artifacts.

How Well Can Short Time-Course Events be Mapped in the Human brain?

Mapping and analysis of short time-course brain events and of temporo-spatial correlations of distributed cerebral activity requires techniques such as electro-encephalography (EEG) which record spontaneous brain activity. Event related magneto-encephalography (MEG) and electrical potential mapping (ERP), in which evoked brain activity is recorded, are also possible. Recording and imaging electrical or magnetic changes is limited by spatial sampling. In the case of electrical activity there is signal attenuation by cranial tissues, and in the case of MEG, by the orientation of the active neuronal groups in cortex relative to the surface of the head. The recordings represent integrals of activity from regions determined by spatial resolution, which itself is a function of the number of recording electrodes. Much previous work with MEG and EEG has used unrealistic approximations of the size and shape of the brain to solve for the origin and distribution of recorded signals. More modern methods, using realistic information from anatomical MRI scans, are improving this aspect of the analysis of MEG- and EEG-based data.

Neuro-Imaging Data Can be Used to Investigate Functional Interactions between Brain Regions

In general, multiple scans recorded in time generate the conditions required for investigation of spatio-temporal correlations of cognitively associated brain activity. Volumes of data can be processes using advanced statistical mapping techniques such as principal components analysis, structural equation models and multivariate regression analysis. These result in the generation of statistical images (known as eigenimages) that can be made to reflect functional and effective connectivity between brain areas and to dissociate independent systems that interact at common sites in the brain (for a review of these techniques see Frackowiak and Friston 1994). Functional connectivity reflects the correlated firing of different cortical areas; it may result from a conjoint drive from an unknown, possibly lower level area and therefore need not reflect the influence of one cortical area on the other. Effective connectivity is equivalent to the intuitive notion of synaptic strength, i.e., the effect one cortical area has on another. The techniques should, in principle, be as applicable to fMRI scans as to the PET scan series on which they were developed. The quantitation of effective connectivity, on a neuronal population basis, presents opportunities for experimentation designed to modulate interactions and strengths of functional connections, a particularily interesting way of monitoring plasticity in the brain.

Can Cognitive Processes be Added and Subtracted as Independent Unitary Processes?

The use of conventional categorical (subtraction) experimental designs for the study of cognition is not always appropriate because of the assumptions of pure insertion and hierarchical organization of cognitive processes (Friston et al. 1996). The Donders paradigm can be used to indicate that one task differs in some way from another, but though this methodology is indicative it is not always sufficient to allow definitive interpretation.

There are processes that are activated by stimuli despite the absence of explicit instructions designed to engage them. If words are presented visually and the explicit task is to attend to orthographic features of the words, rather than to their meaning, a whole language network comprising both posterior and inferior temporal, inferior frontal, lateral premotor, supplementary motor and supramarginal cortices in activated. Such obligatory activation in the absence of explicit instruction or conscious effort may negate a simple experimental approach dependent on additive logic.

Some processes are enhanced by directed attention. Activity in specialised extrastriate areas evoked by passive viewing of identical visual displays can be modulated by drawing attention to one or other attribute or by dividing attention among the attributes (Corbetta et al. 1991). There is, however, little or no information about the effect of unconscious operations on local brain activity.

Neuro-Imaging Studies of the Active Brain Depend on Self-Reporting

Many functional neuro-imaging experiments depend on self-reporting by the experimental subject. This fact is sometimes used to suggest that the scientific data thus obtained are "soft." Introspection and reporting of observations or actions is an intimate part of everyday existence, even in the sphere of science. A statement that an object is a metre long is an example of self-reporting by the measurer. We do not doubt the report because it is consistent, it can be repeated by others and, for example, reliable working machinery can be made based on the correctness of the measurement. Introspection has been used to investigate human brain function by neurologists and neuropsychologists with the clinico-pathological lesion method for at least a century. The repeated demonstration of consistent patterns of local brain activation during defined mental activity now provides objective measurements that are difficult to refute. Such measurements bring the investigation of thought, consciousness, emotion and similar brain functions into the realm of "hard" scientific enquiry.

Cognitive tasks dependent on self-reporting are often associated with responses that can be recorded or physiological changes that can be measured. Parametric scanning experiments that depend on recording task performance and correlating results with scan data are able to identify regions in which activation is altered as a function of a task. Task difficulty itself may be varied. Parametric experiments obviate the need for control scans and potentially circumvent the assumption of pure insertion. For example, in the visual system activity in visual cortex increases with rising flash frequency, reaches an apex and falls off at frequencies that become difficult to discriminate. This result has been replicated using fMRI and PET methods. The perception of pitch can be tested with pure tones of different frequency, a strategy used with MEG to describe tonotopic maps in the transverse temporal gyrus (Sherg and von Cramon 1986). A change in the slope (or intercept) of a task-activity relationship as a result of a modulating influence can be used as a measure of plasticity.

At a more cognitive level, in primary auditory cortex, which is the projection site of the auditory apparatus, activity increases linearly with the number of words spoken per unit time. The posterior temporal cortex, to which auditory cortex projects, shows a different response with such stimuli. Activation is apparent as soon as words are heard, but no further change in local brain activity occurs across the range of word frequencies (Price et al. 1992). The conversion of a time-dependent response in primary auditory cortex to one in which activity is time-independent suggests a possible mechanism for integration of frequency-determined neural activity into a form that signals phonological or semantic identity.

Similar findings have been observed in the motor system. The activity in primary motor cortex, posterior SMA, ventrodorsal cingulate cortex and cerebellar vermis increases exponentially with rate of repetitive movement or degree of constant force exerted. There are no such exponential changes in other motor related areas, for example in the rostral cingulate, premotor and dorsolateral prefrontal cortices (Dettmers et al. 1995).

Sometimes a correlative (parametric) approach to neuro-imaging experiments is particularly informative. The role of medial temporal structures in human memory has been recognised since the description of the amnesic syndrome in patients with bilateral temporal excision or damage. This structure has been difficult to activate by standard subtraction paradigms except by a specific word-stem completion task that has been reported to activate the right parahippocampal gyrus (Squire et al. 1992). On the other hand, the left hippocampal gyrus can be activated very effectively by increasing load on verbal memory and correlating hippocampal activity with performance in an immediate recall task (Grasby et al. 1994).

What is the Significance of Local De-Activations?

It is an empirical fact that significant whole brain (global) average blood flow changes are rare between activated and control scans, especially when tasks are sufficiently specified and close to each other in processing terms. Relative focal activations are usually accompanied by relative deactivations. It may be important to know if a de-activation is relative or absolute. For this reason it is often useful to include a low-level reference scan in an experiment. A de-activation may be due to a real fall of local perfusion in the experimental scan compared to control, or alternatively a less marked increase in the experimental scan than in the control condition. These alternatives can be distinguished by comparison with an independent reference. Parametric designs allow observation of task-variable dependent decreases, as well as increases, in brain activity without a need for specific control or reference states. A less attractive option is to make measurements of absolute blood flow that are only possible, at present, if arterial blood samples are taken.

Plasticity – a Functional System Level Definition

I will define cerebral plasticity in terms of a long-term alteration in patterns of task or behaviour related activity in distributed brain systems. This theme will be developed to show how such a concept and the mechanisms it implies can be investigated and mapped using modern non-invasive functional imaging techniques. I will use examples from normal behaviour especially in relation to memory, learning and the acquisition of motor skills. I will then discuss the functional reorganisation that follows brain injury and that is associated with spontaneous recovery from motor deficits. Finally I will give an example of modification of patterns of activity in a fronto-temporal distributed system associated with word generation that is modified by disease and modulated by dopaminergic manipulation.

Models of Memory and the Localisation of Psychological Processes to Brain Areas

We have, thus far, discussed functional neuro-imaging in brain systems related to input and output systems. The discussion has progressed into the cognitive domain with examples of increasing complexity and an analysis of task-dependent issues relevant to informative experimental design. The next part of this chapter will deal with studies relevant to the functional neuroanatomy of human memory. This topic is characterised by a large and increasing variety of identified memory processes and complex interactions between them. This fact makes particular demands on appropriate choice of imaging method and task definition for the attribution of function to structure and for monitoring plastic change in activity patterns associated with learning.

Working Memory

Many cognitive processes are composed of a number of sub-processes, some of which can be inferred from double dissociations in patients or from chronometric data in normal subjects, and others that may be unsuspected or unknown. The model of verbal working memory developed by Baddeley and coworkers incorporates at least two sub-processes, a rehearsal system and a phonological store (Baddeley et al. 1992). The former refreshes the contents of the latter which acts as a limited size buffer of three to four words with a half-life of approximately two seconds. This conceptual framework has been examined with imaging experiments that have shown a critical role for the inferior frontal lobe for the rehearsal function and also that the inferior parietal lobe is a crucial site for the function of the phonological store. These conclusions can be drawn by controlling for potential confounding sub-processes by the use of a factorial experimental design. The ability to retain a series of letters in working memory can be controlled by a scan in which shapes are remembered that have no phonological connotations during identical conditions of visual presentation. In a second experiment, the ability to make rhyming judgements, a function that, according to the model, does not engage rehearsal, can be controlled by scanning during judgements of shape identity. The interaction between mnestic and phonological dimensions eliminates contributions to the activation pattern from known or unknown interfering processes and can be used to identify areas associated solely with the phonological store (Paulesu et al. 1993).

Episodic memory

Conscious longer term memories acquired as events (episodic memories) present an interesting object of study, in that the process of remembering lists of words can be contaminated by unconscious priming mechanisms that permit recall with above average success, despite the absence of explicit learning. Use has been made of the known interference with acquisition of explicit memories by difficult distracter tasks. Scanning can be carried out during performance of a paired associate word encoding task, or a control repetitive passive listening task with and without concurrent distraction. The passive listening task controls for auditory and other known and unknown components of the task. The interaction of learning and distraction eliminates efficient episodic encoding, thus providing a control activation map representing areas associated with priming and other irrelevant processes. Priming is the well recognised facilitation of recognition caused by prior exposure to a stimulus that was unassociated with an explicit mnestic requirement. The difference between this map and that comparing encoding with control in the non-distracting state indicates areas that are specifically associated with the episodic encoding process. In effect, it is possible to identify in an extensively activated fronto-parietal memory network that encoding is selectively associated with activation of retrosplenial and left dorsolateral prefrontal cortices (Shallice et al. 1994). It is important to remember that, although these areas have a particularly critical importance in normal encoding, memory function remains a property of the whole network.

Another experiment has identified structural correlates of the recall of previously encoded episodic memories. Scanning was performed whilst subjects viewed the first of a previously presented associated pair of words (a category, e.g., country) and asked to attempt recall of the second of the pair (an exemplar, e.g., England). In this task a strategy involving semantic knowledge about the category might contaminate episodic recall, especially if subjects resort to guessing the exemplar if it does not immediately come to mind. Scanning was therefore carried out during a similar task in which a novel, rather than previously encoded, series of category words were presented. This task depends entirely on semantic knowledge and can control for this process in the experimental recall task. A repetition task was used to control out common aspects of listening to words and other less identifiable sub-processes. A comparison of repetition control task scans with experimental scans gives a map of areas involved in both episodic and semantic recall. On the other hand, a comparison of repetition scans with semantic recall scans gives a map of areas specifically associated with semantic recall. The difference between the two isolates areas associated with episodic recall alone (Fletcher et al. 1996). The result, in satisfactory agreement with Tulving's (1983) model, indicates a prominent role for the right prefrontal cortex and also the precuneus in recall of episodic memories.

Plasticity

Motor Associated Cortices are Functionally Specialised and Organised into a Nested Hierarchy of Areas Comprising a Widely Distributed System

Mapping of simple movements about joints in one limb has confirmed a degree of somatotopy with relationships between sites of maximal activation along the motor strip in a disposition approximately illustrated by Penfield's homunculus. The activated areas tend to overlap in extent but the centres of mass of the activations are clearly separated along the central sulcus. Somatotopy, with the arm represented in a dorso-ventral axis and the shoulder lying ventralmost, has also been described in supplementary motor areas on the mesial surface of the cortex (Colebatch et al. 1991).

There are multiple motor representations in human brain that can be demonstrated by selection of appropriate task conditions during scanning. Activations have been found at sites in the insula, in ventral and dorsal premotor cortex, in primary motor and sensory cortices, in rostralmost parts of dorsolateral parietal cortex and in at least three sites in mesial cortex in anterior and posterior SMA and in dorsal and ventral cingulate cortex. There is preliminary evidence that some, if not all, of these areas are somatotopically organised into maps with separate peaks in each activation cluster for movements of different body parts (Fink et al. 1997).

The most extensive activation of motor-related areas is found when executed actions are self-selected, without instruction, from a range of possible directions of movement, for example, in the joystick paradigm of Deiber et al. (1991). In addition to areas already described there is activation of both dorsolateral prefrontal cortices, areas of mesial frontal cortex extending down to the level of the genu of the corpus callosum and into the frontal pole, and areas of lateral and mesial parietal cortex including inferior as well as superior parietal lobules. Activations of the basal structures, especially putamen and thalamus, are most evident when movements are self-paced or when a constant, rather than a phasic, force is exerted (Dettmers et al. 1996). There are major activations of the ipsilateral cerebellar hemisphere and of the cerebellar vermis on movement.

Different Motor Areas are Associated with Different Aspects of Action

Imagining a movement can help improve performance, a fact well known by musicians and athletes. The brain activity associated with imagined actions has been investigated and compared with a resting state, with preparation of motor set and with actual execution of the same movements (Stephan et al. 1995). The network of areas involved in motor imagination surrounds areas that are associated with motor execution. Imagining a complex arm movement is associated with activation of a number of frontal and parietal regions, including bilateral medial and lateral premotor cortices, anterior cingulate cortex, ventral opercular areas and parts of the superior and inferior parietal cortex. Execution of the same complex motor task activates these areas and additionally the contralateral primary sensorimotor cortex and adjacent caudalmost parts of premotor cortex and rostralmost parts of superior parietal cortex. In addition there is activation of parts of cingulate cortex. These areas together form an "executive core" of the motor system.

There is functional specialisation in motor-associated cortices that form recognised anatomical structures. The posterior supplementary motor area (pSMA) (defined as the SMA caudal to the vertical coronal plane at the anterior commissure) can be subdivided functionally into distinct rostral and caudal parts. Rostral pSMA is activated by imagined movement and the more caudoventral pSMA is additionally activated by movement execution. The cingulate functions in a similar rostro-caudally distributed manner. Similar structure-function relationships have been observed by investigators with fMRI (Tyszka et al. 1994).

What Structures Participate in the Initiation of Movement?

Comparison of imagined movements with a resting state, as opposed to preparation-of-set, produces prominent additional activations in the dorsolateral pre-frontal cortex, anterior SMA and anterior cingulate cortex. The role of the pre-frontal cortex in the initiation and selection of movements is substantial (Frith et al. 1991). Controversy exists about the nature of this contribution, especially in view of data suggesting a role for the same area in working memory. Experiments loading short-term memory beyond the span of immediate recall activate DLPFC (Petrides et al. 1993). Other experiments, contrasting intrinsically generated and extrinsically specified movements, specifically activate DLPFC and anterior cingulate (Frith et al. 1991). These latter experiments serve as an example of a scanning paradigm in which cognitive activity is isolated for analysis by keeping constant the sensory and motor components of a task, but varying the nature of the mental operation involved.

Learning New Skills Involves Large Scale Time-Dependent Changes in Patterns of Neuronal Activation

Repetitive task performance results in habituation and adaptation effects that become evident when scans are analyzed. For example, in a verb generation task, responses to novel categories produce activation in a distributed network that includes DLPFC, and in a number of language-related areas (Raichle et al. 1994). If responses are produced to overlearned exemplars, the pattern of activation is attenuated and resembles that obtained with the simple repetition of words, indicating that response selection has become automatic (i.e., performance is accurate whilst another task is being performed).

Motor skill learning has been measured with repetitive performance of a standard pursuit rotor task (Grafton et al. 1992). Improved performance, manifesting as greater accuracy and longer time-on-target, correlates with increased activity in primary and supplementary motor cortices. In this instance a proper interpretation depends on a realisation that performance and skillfulness are separate, but confoundable, attributes of a motor act. If a motor task is performed repeatedly to the same performance criterion, then progressive attenuation of cerebellar and premotor activation is observed, with no change of activity in primary motor cortex (Friston et al. 1992). An experimental design that capitalises on changes of activity in a region with task repetition is very suitable for application with rapid fMRI scanning.

When scanning is performed with error feedback during learning of a novel motor sequence, for example by an auditory signal, there is greater activation in the right premotor area than when the task has been over-learned. Conversely, SMA activity is greater in an automated task than in a naive learning task. With motor learning, visual and language-associated cortices show considerably less activity during learning than when an over-learned key-press sequence is performed (Jenkins et al. 1994; Jueptner et al. 1997). When a task is novel and requires considerable attentional resources, there appear to be mechanisms for large scale de-activation of whole systems that are not required for the task. When a task becomes over-learned, attentional needs decline and activity in extraneous areas normalises relative to activity in the remainder of the brain.

A number of brain areas are associated with divided and selective attention to stimuli. Selective attention, as noted already, results in augmentation of activity in modality-specific areas specialised for the elaboration of a function or percept to which attention is directed. An example from studies of the visual system has already been given. However, there is also additional supramodal activation

of subcortical structures such as the globus pallidus, caudate and posterior thalamus, and of the lateral orbito-frontal and insular/prefrontal cortices. Divided
attention tasks, on the other hand, result in activation of a network consisting of
the anterior cingulate and dorso-lateral pre-frontal cortices (Corbetta et al. 1991).
Attention can modulate activity by altering the gain or responsiveness of an area
to sensory input. However, a second mechanism that increases tonic activity in
an area before expected stimuli arrive has also been described (Rees et al. 1997).

Is there Evidence of Large Scale Plastic Change in Lesioned Brains?

Comparisons of brain activity at rest between patient and control groups can be
informative (Weiller et al. 1992). Disturbances of normal patterns of resting brain
activity caused by internal capsular lesions have been found in both lesioned and
unlesioned hemispheres. In lesioned hemispheres, relative deactivation is
observed in component areas of the cerebral motor system in addition to very
low activity centred on the infarcted internal capsule. These areas include dorsolateral pre-frontal cortex, premotor and parietal cortex, including the insula and
opercular regions, the contralateral cerebellum and nuclei in midbrain and pons
ipsilateral to the lesioned hemisphere. White matter tract degeneration can also
be determined by appropriate anatomical MR scanning with which for example,
degeneration of the pyramidal tract following capsular infarction has been identified in life (Danek et al. 1990; Fries et al. 1991).

   In unlesioned cerebral hemispheres relative hyperactivity is found in posterior cingulate cortex, in ventral premotor cortex and in caudate nucleus. The significance of these unexpected changes is poorly understood. They may be due to
loss of inhibition from the lesioned hemisphere, or to relative augmentation of
activity in cortices that, at least in non-human primates, contain bilateral representations of movement. There is, therefore, profound plasticity of resting state
activity in both hemispheres associated with recovered motor function. Furthermore, the plastic changes occur in brain areas that have anatomical connections
to the lesion or constitute components of the lesioned motor system.

Motor Recovery after Internal Capsule Lesions is Common –
What is the Pathophysiological Basis?

Studies in this area are few. Some degree of spontaneous recovery accompanies
many cases of motor paralysis caused by stroke. Recovery can sometimes be
complete, but its biological basis is largely unknown. Experimental design is
complicated by considerations similar to those discussed in relation to defining
areas of brain associated with learning, practice and performance. It is difficult
to envisage what information about the cerebral basis for motor recovery can be
obtained by comparing task-associated distributions of brain activity in normal
subjects with those obtained in patients who cannot perform an activating task

properly. Studies have therefore been undertaken in completely recovered patients because the criteria of motor task performance can be equalised across patient and control groups.

Comparisons have been made of brain activity associated with distal, fractionated, repetitive opposition movements of fingers and thumb to brain activity at rest using unaffected and recovered hands separately (Chollet et al. 1991). The normal lateralised pattern of cortical, subcortical and cerebellar activity becomes bilateral. Increased activation restricted to motor associated brain areas is observed when a recovered limb moves. Among those regions activated are areas that usually participate in freely selected, complex, motor sequence tasks but not in simple repetitive tasks such as those used by recovered patients in the experiments. Homologous activation of M1 contralateral to a recovered limb is invariably associated with mirror movements of the unaffected limb, making it impossible to decide whether M1 activation represents an epiphenomenon or a causal component of normal spontaneous recovery (Weiller et al. 1992). In summary, changes in both resting and activated brain function are remarkable for their extent and are witness to considerable plastic change in the lesioned brain.

Extension of these studies to people with partial recovery has been difficult. The main problem is confounding effects of differential performance, so studies have been few and preliminary (Dettmers et al. 1997). Preliminary, unpublished evidence, also from our laboratory, suggests that there may be profound effects on the capacity to recover if areas involved in motor initiation, planning and sequencing are dysfunctional. Poor recovery may occur despite activation of motor-sensory cortex if pre-executive motor cortices cannot be activated.

The process of recovery in patients with capsular infarcts may be associated with differential effects on motor cortical (M1) activation (Weiller et al. 1993). In certain circumstances, recovery is accompanied by little change in the extent of activated motor cortex. In other circumstances, the motor representation of finger and thumb movements is greatly enlarged, spreading ventrally onto cortex associated with the face and rostro-caudally into premotor and parietal cortex. The differential activation is associated with a different anatomical locations of lesions in subcortical white matter (Fries et al. 1993). Lesions in the posterior portion of the posterior limb of the internal capsule which interrupt output from M1, result in enlargement of the motor representation. It may be that degeneration of pyramidal neurons in M1, resulting from interruption of axons in the internal capsule, results in a loss of recurrent, collateral, GABA-mediated inhibition of surrounding pyramidal cells (Gosh and Porter 1988). Lesions in the anterior limb produce no such effect and result in augmented activation in premotor and prefrontal areas, depending on the site of the capsular lesion and the provenance of damaged axons.

Similar changes in extent of activation have been found in other pyramidal disorders, e.g., motor neuron disease (Kew et al. 1994). The effect of sensory stimulation or the results of physiotherapy in modifying patterns of activity related to recovery have not been systematically examined. Enlargement of sensory fields in patients with traumatic (as opposed to congenital) amputation of an upper limb has been described (Kew et al. 1994). There are also marked changes in motor cortical activation produced by stump and contralateral arm movements which are associated with increased cortical excitability to electro-physiological or transcranial magnetic stimulation.

Can Neuromodulatory Effects Responsible for Plasticity Be Monitored in the Human Brain?

There are two principal modes of neurotransmitter action in the central nervous system. One involves fast acting excitatory or inhibitory mechanisms, usually associated with neurotransmitters such as glutamate, aspartate and GABA. A second type operates on a slower time scale and has a primary neuromodulatory mode of action that alters target neuronal excitability. These neuromodulatory systems are further characterised by relatively low baseline firing rates and a prolonged duration of action following neurotransmitter release. Typical neuromodulatory neuro-transmitters are noradrenaline (NA), dopamine (DA) and serotonin (5HT). Fast acting neurotransmitters are implicated primarily in signal transfer; slow acting neurotransmitters alter the excitability of cortical neurones to other extrinsic influences with relatively little effect on background resting potentials.

A critical issue in developing an understanding of human brain function is to establish how modulatory neurotransmitters influence human cortical activity. Experimental protocols using factorial designs, in which the effect of drug can be determined in the presence and absence of some cognitive task, enable these issues to be addressed in the intact human brain (Frackowiak and Friston 1994).

Can Abnormal Cortical Activity in Schizophrenia Be Corrected by Manipulating Dopaminergic Transmission?

Dysfunction of the mesolimbic-mesocortical dopamine system that originates in the ventral tegmentum is thought to be important in the pathophysiology of schizophrenia (Snyder 1973; Stevens 1973). Preliminary findings have suggested significant neuromodulatory effects of dopamine manipulation, with apomor-phine or amphetamine, on the function of the dorsolateral prefrontal cortex in patients with schizophrenia (Daniel et al. 1991). To determine whether there are abnormal neuromodulatory effects of dopamine in schizophrenia we examined its regulatory role on cortical function in a placebo-controlled experiment of unmedicated acute schizophrenics. The effect of a dopaminergic perturbation with apomorphine of a cognitive task-induced cortical activation was directly compared in schizophrenics and normal people (Dolan et al. 1995).

In normal subjects the left dorsolateral prefrontal cortex, the thalamus and the anterior cingulate cortex were activated and the left temporal plane was deac-tivated by the comparison of a repetition with a word fluency task. The same cog-nitive activation in schizophrenic patients resulted in a failure of activation of anterior cingulate cortex and a failure of deactivation of the planum temporale.

Following dopaminergic manipulation the relative failure of task-induced cingulate activation in schizophrenic patients was reversed, as was the activation of the temporal plane. In the presence of a dopaminergic neuromodulatory influence schizophrenic patients displayed a pattern of activation in the word generation associated cortical system that became like that activated in normal subjects. There was no direct effect of apomorphine at the anterior cingulate. Thus, in acute schizophrenia there is a significant effect of central dopaminergic manipulation on a pattern of task-specific neuronal responses in anterior cingulate cortex. The result provides direct evidence of abnormal neuromodulatory effects of dopamine on prefrontal function in schizophrenia. In this model the effect of apomorphine is primarily presynaptic so that there is a net decrease in dopaminergic neurotransmission and enhanced responses in cingulate cortex because of release from a tonic inhibitory dopamine input.

Conclusions

The field of functional brain mapping is in a state of technical development. The data derived from the different mapping modalities are often complementary, and there is much evidence to suggest that this state of affairs will continue, thereby providing improved tools for the exploration of the functional architecture of the human brain. Neuroscientific progress will be made by judicious use of one or more methods to answer the appropriate question (Buchner et al. 1994). Advances in fMRI and in electrophysiological modalities promise a much needed increase in spatial and temporal resolution. PET studies have provided much of the initial functional anatomic data from the human brain. Many results, though preliminary, indicate avenues of further enquiry. We can conclude, on the basis of the results, that it is now possible to embark on research into the functional architecture of the living human brain that goes beyond descriptive "neophrenology." Nevertheless, it remains a fact that our description of the anatomical arrangement of the functioning human brain is incomplete and therefore even "neophrenology" remains an important area of continuing study. Uniquely human brain functions need assignment to networks of brain areas and cognitive processes require definition in physiological and anatomical terms. New abilities, such as the measurement of strengths of functional connections between brain areas (Friston et al. 1993), may have practical significance for treatment and modification of disease and the understanding of memory and learning.

References

Baddeley AD (1992) Working memory. Science 255:556–559

Buchner H, Weyen U, Frackowiak RSJ, Romaya J, Zeki S (1994) The timing of activity in human area V4. Proc R Soc Lond B: 257:99–104

Chollet F, DiPiero V, Wise RJS, Brooks DJ, Dolan RJ, Frackowiak RSJ (1991) The functional anatomy of motor recovery after ischaemic stroke in man. Ann Neurol 29:63–71

Colebatch JG, Deiber MP, Passingham RE, Friston KJ, Frackowiak RSJ (1991) Regional cerebral blood flow during voluntary arm and hand movements in human subjects. J Neurophysiol 65:1392–1401

Corbetta M, Miezin FM, Dobmeyer S, Shulman GL, Petersen SE (1991) Selective and divided attention during visual discriminations of shape, color, and speed: Functional anatomy by positron emission tomography. J Neurosci 11:2383–2402

Danek A, Bauer M, Fries W (1990) Tracing of neural connections in the human brain by magnetic resonance imaging in vivo. Eur J Neurosci 2:112–115

Daniel DG, Weinberger DR, Jones DW, Zigun JR, Coppda R, Handel S, Bigelow LB, Goldberg TE, Berman KF, Kleinman JE (1991) The effect of amphetamine on regional cerebral blood flow during cognitive activation in schizophrenia. J Neurosci 11:1907–1917

Deiber MP, Passingham RE, Colebatch JG, Friston KJ, Nixon PD, Frackowiak RSJ (1991) Cortical areas and the selection of movement: a study with PET. Exp Brain Res 84:392–402

Dettmers C, Fink GR, Lemon RN, Stephan KM, Passingham RE, Silbersweig D, Holmes A, Ridding MC, Brooks DJ, Frackowiak RSJ (1995) Relation between cerebral activity and force in the motor areas of the human brain. J Neurophysiol 74:802–815

Dettmers C, Lemon RN, Stephan KM, Fink GR, Frackowiak RSJ (1996) Cerebral activation during the exertion of sustained static force in man. NeuroReport 7:2103–2110

Dettmers C, Stephan KM, Lemon RM, Frackowiak RSJ (1997) Reorganization of the executive motor system after stroke. Cerebrovasc Dis 7:187–200

Dolan RJ, Fletcher P, Frith CD, Friston KJ, Frackowiak RSJ, Grasby PM (1995) Dopaminergic modulation of impaired cognitive activation in the anterior cingulate cortex in schizophrenia. Nature 378:180–182

Fink GR, Frackowiak RSJ, Pietrzyk U, Passingham RE (1997) Multiple non-primary motor areas in the human cortex. J Neurophysiol 77:2164–2174

Fletcher PC, Shallice T, Frith CD, Frackowiak RSJ, Dolan RJ (1996) Brain activity during memory retrieval. The influence of imagery and semantic cueing. Brain 119:1587–1596

Frackowiak RSJ, Friston KJ (1994) Functional neuroanatomy of the human brain: positron emission tomography – a new neuroanatomical technique. J Anatomy 184:211–225

Frackowiak RSJ, Frith CD, Dolan RJ, Mazziota JC (1997) Human brain functions. Academic Press, San Diego

Fries W, Danek A, Witt TN (1991) Motor responses after transcranial electrical stimulation of cerebral hemispheres with a degenerated pyramidal tract. Ann Neurol 29:646–650

Fries W, Danek A, Scheidtmann K, Hamburger C (1993) Motor recovery following capsular stroke – role of descending pathways from multiple motor areas. Brain 116:369–382

Friston KJ, Frith CD, Passingham RD, Liddle PF, Frackowiak RSJ (1992) Motor practice and neurophysiological adaptation in the cerebellum: a PET study. Proc R Soc Lond Biol 243:223–228

Friston KJ, Frith CD, Frackowiak RSJ (1993) Time dependent changes in effective connectivity measured with PET. Human Brain Mapping 1:69–80

Friston KJ, Price CJ, Fletcher P, Moore C, Frackowiak RSJ, Dolan RJ (1996) The trouble with cognitive subtraction. NeuroImage 4:97–104

Frith CD, Friston KJ, Liddle PF, Frackowiak RSJ (1991) Willed action and the prefrontal cortex in man. Proc R Soc Lond B 244:241–246

Gosh S, Porter R (1988) Morphology of pyramidal neurons in monkey motor cortex and the synaptic actions of their intracortical axon collaterals. J Physiol 400:593–615

Grafton S, Mazziotta JC, Presty S, Friston KJ, Frackowiak RSJ, Phelps ME (1992) Functional anatomy of human procedural learning determined with regional cerebral blood flow and PET. J Neurosci 12:2542–2548

Grasby PM, Frith CD, Friston KJ, Simpson J, Fletcher PC, Frackowiak RSJ, Dolan RJ (1994) A graded task approach to the functional mapping of brain areas implicated in auditory-verbal memory. Brain 117:1271–1282

Jenkins IH, Brooks DJ, Nixon PD, Frackowiak RSJ, Passingham RE (1994) Motor sequence learning: a study with positron emission tomography. J Neurosci 14:3775–3790

Jueptner M, Frith CD, Brooks DJ, Frackowiak RSJ, Passingham RE (1997) Anatomy of motor learning. 2. Subcortical structures and learning by trial and error. J Neurophysiol 77:1325–1337

Kew JJM, Ridding MC, Rothwell JC, Passingham RE, Leigh PN, Sooriakumaran S, Frackowiak RSJ, Brooks DJ (1994) Reorganisation of cortical blood flow and transcranial magnetic stimulation maps in human subjects after upper limb amputation. J Neurophysiol 72:2517–2524

Paulesu E, Frith CD, Frackowiak RSJ (1993) Functional anatomy of the articulatory loop. Nature 362:342–345

Petrides M, Alivisatos B, Meyer E, Evans AC (1993) Functional activation in the human frontal cortex during the performance of verbal working memory tasks. Proc Natl Acad Sci USA 90:878–882

Price C, Wise RJS, Ramsay S, Friston KJ, Howard D, Patterson K, Frackowiak RSJ (1992) Regional response differences within the human auditory cortex when listening to words. Neurosci Lett 146:179–182

Raichle ME, Fiez JA, Videen TO, Macleod AK, Pardo JV, Fox PT, Petersen SE (1994) Practice-related changes in human brain functional anatomy during non-motor learning. Cereb Cortex 4:8–26

Rees G, Frackowiak RSJ, Frith CD (1997) Two modulatory effects of attention that mediate object categorization in human cortex. Science 275:835–838

Shallice T, Fletcher P, Frith CD, Grasby P, Frackowiak RSJ, Dolan RJ (1994) The brain regions associated with the acquisition and retrieval of verbal episodic memory. Nature 368:633–635

Sherg M, von Cramon D (1986) Evoked dipole source potentials in the human auditory cortex. Electroencephalogr Clin Neurophysiol 65:344–360

Snyder SH (1973) Amphetamine psychosis: a model schizophrenia mediated by cathecolamines. Am J Psychiat 130:61–67

Squire LR, Ojemann JG, Miezin FM, Petersen SE, Videen TO, Raichle ME (1992) Activation of the hippocampus in normal humans: a functional anatomical study of memory. Proc Natl Acad Sci USA 89:1837–1841

Stephan KM, Fink GR, Passingham RE, Silbersweig D, Ceballos Baumann A, Frith CD, Frackowiak RSJ (1995) Imaging the execution of movements: functional anatomy of the mental representation of upper extremity movements in healthy subjects. J Neurophysiol, 73:373–386

Stevens JR (1973) An anatomy of schizophrenia? Arch Gen Psychiatr 29:177–189

Tulving E (1983) Elements of episodic memory. Oxford University Press, Oxford

Tyszka JM, Grafton ST, Chew W, Woods RP, Colletti PM (1994) Parcellation of mesial frontal motor areas during ideation and movement using functional magnetic resonance imaging at 1.5 tesla. Ann Neurol 35:746–749

Weiller C, Chollet F, Friston KJ, Wise RJS, Frackowiak RSJ (1992) Functional reorganisation of the brain in recovery from striatocapsular infarction in man. Ann Neurol 31:463–472

Weiller C, Ramsay SC, Wise RJS, Friston KJ, Frackowiak RSJ (1993) Individual patterns of functional reorganisation in the human cerebral cortex after capsular infarction. Ann Neurol 33:181–189

Connectionist Modeling of Relearning and Generalization in Acquired Dyslexic Patients

D. C. Plaut*

Summary

Connectionist models implement cognitive processes in terms of cooperative and competitive interactions among large numbers of simple, neuron-like processing units. Such models provide a useful computational framework in which to explore the nature of normal and impaired cognitive processes. The current work extends the relevance of connectionist modeling in neuropsychology to address issues in cognitive rehabilitation: the degree and speed of recovery through retraining, the extent to which improvement on treated items generalizes to untreated items, and how treated items are selected to maximize this generalization. A network previously shown to model impairments in mapping orthography to semantics was retrained after damage. The degree of relearning and generalization depended on the location of the lesion and had interesting implications for understanding the nature and variability of recovery in patients. In a second simulation, retraining on words whose semantics are atypical of their category yielded more generalization than retraining on more typical words, suggesting a counterintuitive strategy for selecting items in patient therapy to maximize recovery. Taken together, the findings demonstrate that the nature of relearning in damaged connectionist networks can make important contributions to a theory of rehabilitation in patients.

Introduction

It was once thought that the brain lost much of its plasticity beyond an early, *critical period* in development. There is now, however, considerable evidence that the response properties of cortical neurons in adult animals can be remapped extensively in response to intensive training regimes (see Kaas 1994; Merzenich and Jenkins 1995 for reviews). Although most of these demonstrations have been in the peripheral domains of sensory and motor processing, more recent attempts to apply analogous intervention strategies to language-learning impaired children (Tallal et al. 1993) have met with considerable success (Merzenich et al.

* Departments of Psychology and Computer Science and the Center for the Neural Basis of Cognition, Carnegie Mellon University

J. Grafman / Y. Christen (Eds.)
Neuronal Plasticity:
Building a Bridge from the Laboratory to the Clinic
© Springer-Verlag Berlin Heidelberg New York 1999

1996; Tallal et al 1996). Among other things, these findings raise considerable hope for the development of more effective strategies for remediating the cognitive impairments of individuals with brain damage.

There remains, however, something of a puzzle. If the brain remains so plastic well into adulthood, why is the prognosis for recovery of cognitive functions following brain damage often so poor? While patients with certain types of brain damage may show nearly complete post-morbid recovery (e.g., patients with hemispatial neglect following right parietal damage; see Robertson and Marshall 1993), others, particularly those with language impairments, do not (see Kertesz 1985). Moreover, even in the circumstances in which there is substantial recovery of function, little is known about its physiological and cognitive bases, or what factors might influence its effectiveness (Hillis 1993).

The current work attempts to provide a theoretical framework, supported by explicit computational simulations, for understanding how and when experience-driven therapy is most effective at remediating cognitive impairments. The focus is not only on the degree and speed with which behavior can be reestablished as a result of therapy, but also on the extent that recovery due to treatment of particular items generalizes to other materials, and the possible bases on which to select items for treatment so as to maximize this generalization. The work is cast in terms of connectionist or parallel distributed processing models, in which information is represented as patterns of activity over large groups of simple, neuron-like units. Processing takes the form of cooperative and competitive interactions among the units on the basis of weighted connections between them. These weights encode the long-term knowledge of the system and are learned gradually through experience in the domain. Models of this form can be developed within a wide range of cognitive domains, including high-level vision and attention, learning and memory, speech and language processing, and the coordination and control of action (see McClelland et al. 1986; Qinlan 1991; Rumelhart et al. 1986b).

The effects of damage in connectionist models have been used to account for a number of specific neuropsychological disorders. Brain damage can be approximated within such models by the removal of some proportion of the units and/ or connections in certain regions of the model. Perhaps the most widely investigated class of disorders concern selective impairments in reading the acquired dyslexias (Hinton and Shallice 1991; Mozer and Behrmann 1990; Plaut and Shallice 1993; Plaut et al. 1996). The current work extends the relevance of connectionist modeling in cognitive neuropsychology by demonstrating that the same computational principles that are effective for understanding normal cognitive processing, and the effects of brain damage, can also provide insight into the nature of recovery from brain damage.

The next section provides a brief overview of findings from empirical studies attempting to remediate the reading impairments of acquired dyslexic patients. Following this, two computational simulations are presented, both involving networks that are trained to derive the meanings of written words (see Plaut 1996 for more details). The first demonstrated that, in retraining a network after damage, the degree of relearning and generalization depended on the location of the lesion. The results have interesting implications for understanding the nature and variability of recovery in patients. In the second simulation, retraining on words whose semantics are atypical of their category yielded more generalization than retraining on more typical words, suggesting a counterintuitive strategy for selecting items in patient therapy to maximize recovery. Taken together, the findings demonstrate that the nature of relearning in damaged connectionist networks can make important contributions to a theory of rehabilitation in patients.

Remediation of Acquired Dyslexia

Coltheart and Byng (1989) undertook a series of remediation studies with a surface dyslexic patient, EE, with left temporal-parietal damage due to a fall. On the basis of a number of preliminary tests, Coltheart and Byng determined that EE had a specific deficit in deriving semantics from orthography. In one study, they gave EE 485 high-frequency words for oral reading. The 54 words he misread were divided in half randomly into treated and untreated sets. For words in the treated set, EE studied cards of the written words augmented with mnemonics for their meanings. As a result, his reading performance on the treated words improved from 44 % to 100 % correct. Surprisingly, the untreated words also improved, from 44 % to 85 % correct. This improvement was not due to "spontaneous recovery" or to other non-specific effects because performance on the words was stable both before and after therapy. Two other studies with EE produced broadly similar results. Overall, Coltheart and Byng found excellent recovery of treated items and substantial generalization to untreated items (also see Weekes and Coltheart 1996).

A useful measure of generalization is the amount that untreated items improve relative to the the amount that they would have improved if they had been treated directly. This measure can be approximated by the ratio of the improvement on untreated items to the improvement on the treated items. Thus, Coltheart and Byng's (1989) therapy with EE produced 41/56 = 73 % generalization.

Unfortunately, such promising results are not always found in rehabilitation studies, even those with very similar types of patients. Scott and Byng (1989) treated a surface dyslexic patient for homophone confusions in reading (e.g., TAIL/TALE) and produced improvement on treated items and, to a lesser extent, untreated items, but found no generalization to his writing of the same items (also see Behrmann 1987). Behrmann and Lieberthal (1989) trained a globally aphasic patient with semantic impairments on a semantic category sorting task. They found improvement on untreated items only within some categories and minimal generalization to items in untreated categories. Finally, Hillis (1993) carried out an extensive rehabilitation program with a patient who had both orthographic and semantic impairments. The patient was able to learn trained tasks (e.g., lexical decision, naming) but showed virtually no generalization to untrained tasks.

Why some patients improve while others do not is not entirely clear. Furthermore, even in those patients who do improve and show generalization, the cause of this generalization – in terms of changes to the underlying cognitive mechanism induced by treatment – is unknown. An explanation of these findings should account not only for the occurrence of generalization in some patients and conditions, but also for its absence in others. As Hillis (1993) points out, what is needed is a theory of rehabilitation that provides a detailed specification of the impaired cognitive system, how it changes in response to treatment, and what factors are relevant to the efficacy of the treatment.

A Connectionist Approach to Remediation

Early connectionist research (Hinton and Plaut 1987; Hinton and Sejnowski 1986) demonstrated that simple networks trained on unstructured tasks can, when retrained after damage, exhibit rapid recovery on treated items and generalization to untreated items. Plaut (1996) extended these findings to apply directly to understanding the basis and variability of recovery in patients, and to provide a platform for testing hypotheses on how to select items for treatment to maximize generalized recovery. The modeling work was cast within a more general framework of lexical processing (see Seidenberg and McClelland 1989; Plaut et al. 1996) in which distributed representations of written words (orthography), spoken words (phonology) and their meanings (semantics) interact to simultaneously settle on the best interpretation of an input (see Fig. 1).

Simulation 1: Recovery and Generalization

The network used in the first simulation, depicted in Figure 2, was based closely on the one used by Hinton and Shallice (1991) and constitutes an implementation of the orthography-to-semantics portion of the general framework in Figure 1. Written input, in the form of 40 four-letter words, was presented to the network by clamping particular patterns of activity (1s and 0s) over 32 *orthographic* units (eight features per letter). The meaning of each of the words, falling into five categories of eight items each, was represented by a particular pattern of activity over

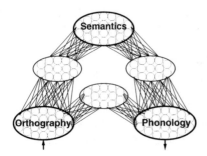

Fig. 1. A connectionist framework for lexical processing. (Adapted from Plaut 1997)

68 *semantic* units. The assignment of semantic features to words had the property that words within the same category tended to have more similar semantic representations than those in different categories, and there was no systematic relationship between orthographic similarity and semantic similarity (see Plaut and Shallice 1993 for details). During processing, the activation levels of units were a smooth, nonlinear (sigmoidal) function of their summed weighted inputs from other units, ranging between 0 and 1. The network was trained with a version of back-propagation appropriate for recurrent networks, known as *back-propagation through time* (Rumelhart et al. 1986a), to activate the appropriate semantic features of a word when presented with its orthographic representation. In doing so, the network learned to use bidirectional interactions between the semantic units and so-called *clean-up* units to make the meanings of words into stable, *attractor* patterns.

Once the network had learned to derive the meanings of the 40 words from their written form accurately, it was damaged in one of two places: near orthography (the connections from the orthographic units to the intermediate units) or near semantics (the connections from the clean-up units to the semantic units). These two locations are indicted in bold in Figure 2. Lesions involved removing a random proportion of the indicated set of connections; the severity of the lesion was controlled by changing the proportion of connections that were removed.

After a given instance of lesion, the network was presented with each of the 40 words for processing. As a result of the damage, the semantic activity produced by the network would often differ significantly from the correct semantics of the presented word. The network was considered to have responded correctly if the *proximity* (i.e., normalized dot-product) of the semantics generated by the network was within 0.8 of the correct semantics of the presented word, and the proximity of the next best word was at least 0.05 further (see Hinton and Shallice 1991 for details). If the generated semantics satisfied these criteria when compared with the semantics of some word other than the one presented, that word was considered to be the network's response (an error). Otherwise, the network was considered to have failed to respond (an omission). The response criteria can be thought of as substituting for the semantics-to-phonology portion of the general framework (see Plaut and Shallice 1993 for implementations).

For each of the two locations of damage, a severity of lesion was chosen that reduced correct performance on the 40 trained words to 20%. This turned out to involve removing 30% of connections for the orthographic lesions and 50% of connections for the semantic lesions.

Fig. 2. The connectionist network used in Simulation 1. Arrows represent 25% connectivity between groups of units. The simulation contrasts the effects of lesions to the two sets of connections shown in bold. (Adapted from Plaut 1996)

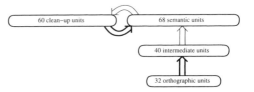

Once the performance of the lesioned network on all 40 words was determined, half of the correct words and half of the incorrect words were randomly selected and placed in the *treated* set; the remaining words were placed in the *untreated* set. Thus, both the treated and untreated sets contained 20 words and were balanced for correct performance. For the purpose of setting up the treated and untreated sets, explicit errors and omissions were both considered incorrect and were not distinguished. The lesioned network was then retrained for 50 epochs (training presentations) on the treated words only. Performance was measured at each epoch during relearning separately for the treated and untreated word sets, in terms of the number of words read correctly, and the average proximity of the generated and correct semantics. To ensure that any relearning effects were not simply due to an imbalance in initial performance between the treated and untreated sets, the two sets were exchanged and the retraining was repeated, starting from the same initial set of weights. Thus, each group of words served both as the treated set and the untreated set. Finally, for purposes of comparison, the weights were again reinitialized and the lesioned network was retrained on all 40 words. Results were averaged over 20 instances of each location and severity of lesion, in which a different random subset of connections were removed.

Figure 3 shows the improvement in performance during retraining following lesions near semantics versus lesions near orthography. Considering the former first (see Fig. 3 a), the network shows rapid relearning of the 20 treated words after a semantic lesions, reaching near perfect performance (98.4 % correct) after 50 training presentations. Moreover, performance on the 20 untreated words in this condition also improved considerably, reaching 67.6 % correct at this point. The improvement on untreated words was 61 % as large as the improvement on the treated words themselves, which is comparable to the 73 % generalization found by Coltheart and Byng (1989) for patient EE.

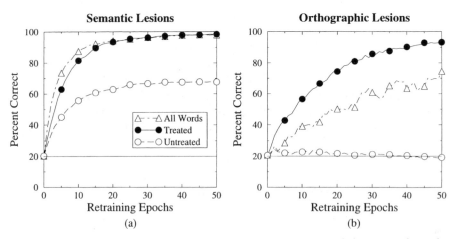

Fig. 3. Improvement on treated and untreated items when retraining a network that maps orthography to semantics after 1) lesions near semantics or 2) lesions near orthography. (Adapted from Plaut 1996)

By contrast, retraining after a lesion near orthography produces quite different results (see Fig. 3 b). In particular, recovery of the 20 treated items is far less effective, although it does reach 93.3 % correct after 50 epochs. More critically, performance on the untreated words fails to improve at all, remaining at near 20 % correct. Thus, the network exhibited poorer recovery and no generalization following retraining after orthographic lesions.

A clue to the basis for this difference can be found from examining the lesioned network's performance when retrained on all 40 words. Following semantic lesions, recovery is faster when retraining on all 40 words than when retraining on only 20 words. By contrast, following orthographic lesions, retraining on all 40 words produces much poorer recovery than when retraining on only 20 words. These findings make sense if the weight changes induced by retrained words after semantic lesions are more consistent across words than after orthographic lesions. The actual weight changes administered to the network after a retraining epoch are the sum of the weight changes induced by each individual word (scaled by the learning rate). Weight changes that are consistent across retrained words accumulate, resulting in fast learning; weight changes that are inconsistent cancel each other out, resulting in much slower learning. Figure 4 presents a graphical depiction of this effect using vectors (arrows) to represent weight changes. Within semantics, similar words require similar interactions, so that the weight changes caused by retraining on some words will tend also to improve performance on other, related words (i. e., the optimal weight changes for words are mutually consistent). By contrast, similar orthographic patterns typically must generate very different semantic patterns. As a result, when retraining after lesions near orthography, the weight changes for treated items are unrelated to those that would improve the untreated items, and there is no generalization.

As reviewed earlier, studies of cognitive rehabilitation of acquired dyslexics have demonstrated considerable relearning of treated items and (often) improvement on untreated but related items. At a general level, the cause of rapid relearning and generalization in the network may provide an explanation for the nature of recovery in these patients. At a more specific level, the finding that the extent

Fig. 4. Depiction of the effect of consistent vs. inconsistent weight changes on the extent of recovery and generalization in relearning. In each condition, the small solid arrows represent directions of weight change induced by treated words; the large solid arrow is the (vector) sum of these smaller arrows, representing the actual weight changes administered to the network. The length of this vector reflects the speed of relearning the treated words. The dotted arrows represent directions of weight change that would be optimal four untreated words if they were trained; to the extent that these point in the same direction as the actual weight change vector, retraining on the treated words will also improve performance on the untreated words. (Adapted from Plaut 1996)

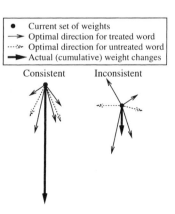

of relearning depends on the location of damage may provide an explanation for why only some patients show substantial recovery and generalization. The simulation results suggest that a patient with a functional impairment close to or within semantics should show considerable generalization, while one with an impairment close to orthography should show little or none. Conversely, the degree of generalization observed in a patient can be used to predict the fine-grained location of their functional impairment *within* the semantic route.

Simulation 2: Designing Retraining to Maximize Generalization

Ideally, an understanding of the impairment in a particular patient should lead to the design of a rehabilitation strategy that maximizes recovery. A potential benefit of connectionist modeling in neuropsychological rehabilitation is that it can provide a framework for investigating the relative effectiveness of alternative rehabilitation strategies. One aspect of a retraining simulation that is under experimental control, and that might influence the nature of recovery, is the selection of items for treatment.

An important aspect of the structure of semantic representations, at least of nouns, is that they are organized into categories. Furthermore, relative to this category structure, a critical semantic variable is typicality – how close a concept is to the central tendency of its category (Rosch 1975). For instance, a robin is highly typical among birds, an eagle is less typical, and a penguin is highly atypical. The question is, is it better to retrain on typical or atypical words? A natural intuition is that relearning the central tendency of a category – that is, retraining on typical words – should lead to the greatest generalization to other words in the category. The results of the current simulation, however, show the opposite: retraining on words that are somewhat atypical of their semantic category leads to greater generalization than retraining on more typical words. The reason, put briefly, is that atypical words collectively convey more information on the overall structure of the category – specifically, on how semantic properties can *vary* across category members – while still providing a good approximation of the central tendency of the category.

The simulation used an artificial version of the task of mapping orthography to semantics in order to more carefully control the nature of the semantic categories. The training set consists of 100 artificial "words." The orthographic representation of each word was created by randomly assigning it an average of 4 out of 20 possible orthographic features. The semantic representations of words were generated by distorting a "prototype" pattern that was generated randomly to have 10 of 50 possible semantic features. The degree of typicality of words is reflected in the number of features in which its representation differs from the prototype: typical words share most of the features of the prototype, while atypical words share far fewer. To implement this, two sets of 50 word meanings were generated from the prototype using different levels of random distortion. The *typical* set consisted of instances produced by a small distortion of the prototype;

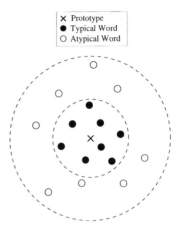

Fig. 5. A depiction of the relationship in semantic space between the prototype of a category and typical versus atypical exemplars in that category. (Adapted from Plaut 1996)

each semantic feature had a probability $d = 0.1$ of being randomly regenerated (with probability $p = 0.2$). The *atypical* set consisted of instances generated using a large distortion ($d = 0.5$). Geometrically, if the prototype corresponds to a particular point in the space of semantic representations, the typical words are points that are near the prototype, while the atypical words are farther away (see Fig. 5). Orthographic patterns were assigned to semantic patterns randomly to ensure that, as in English, there was no systematic relationship between orthography and semantics.

A network was trained with back-propagation through time to activate the correct semantic pattern for each of the 100 artificial words when presented with its orthographic representation. The architecture of the network was broadly the same as the network from the first simulation (see Fig. 2). The differences were that, in the current network, there were only 20 orthographic units (compared with 32), 50 semantic units (compared with 68), and 50 % of the possible connections between groups of units are included (compared with 25 %).

After training, the network was lesioned by removing a randomly selected 25 % of the connections between the intermediate units and the semantic units. This lesion location and severity was selected because it produces an intermediate amount generalization (27 %; see Plaut 1996), providing a clear opportunity for the composition of the treated set to have either a positive or negative impact on generalization.

After each lesion, performance on all 100 words was measured. A presented word was considered correct if the semantics generated by the network had a higher proximity (normalized dot product) to the correct semantics for the word than to the semantics for any other word. Based on this initial performance, the typical and atypical word sets each were divided in half, balancing for correct performance. The lesioned network was then retrained for 50 epochs, either on half of the typical words or on half of the atypical words (25 words). During retraining, improvement in correct performance was measured on this treated set as well as on two untreated sets: the remaining words of the same type (typi-

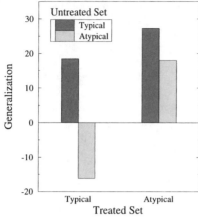

Fig. 6. Generalization from retraining after lesions of 25 % of the intermediate-to-semantics connections, as a function of the semantic typicality of the treated and untreated sets. (Adapted from Plaut 1996)

cal or atypical), and all of the words of the other type. Each half of each group in turn served as the treated set for retraining (reinitializing the weight each time). In this way, the retraining procedure was able to measure the generalization to typical and atypical words when retraining on typical or atypical words.

Somewhat surprisingly, although retraining on typical exemplars produced greater recovery on treated items, retraining on atypical exemplars produced greater generalization to untreated items (see Fig. 6). These findings make sense given the adequacy with which sets of typical versus atypical exemplars approximate the range of semantic similarity among all of the words. Semantically typical words accurately estimate the central tendency of a category, but provide little information about the ways in which category members can *vary*. By contrast, each atypical word indicates many more ways in which members can differ from the prototype and yet still belong to the category. Thus, collectively, the semantic representations of atypical words cover more of the features needed by the entire set of words than do the representations of more typical words. At the same time, the average effects of retraining on atypical words provides a reasonable estimate of the central tendency of the category, yielding generalization to typical words (as found in human category learning by, e. g., Posner and Keele 1968). In this way, the simulation generated a novel prediction about how to select items for treatment so as to maximize generalized recovery.

Conclusion

Attempts at cognitive rehabilitation of acquired dyslexic patients have resulted in considerable improvement in performance on treated words, as well as significant generalization to untreated but related words, although the degree of recovery across patients can vary considerably. There is, however, little understanding of the underlying mechanisms by which cognitive functions recover, either spontaneously or as a direct result of therapeutic intervention. Generalization in the domain of reading via meaning is particularly puzzling as there is no systematic relationship between the written or spoken forms of words and their meanings.

Connectionist modeling offers specific hypotheses about the nature of the representations and computations that underlie cognitive processes, as well as how these processes are learned through experience and how they are affected by brain damage. The current work attempts to extend the relevance of connectionist modeling in neuropsychology one step further, to contribute to a theory of rehabilitation (see Hillis 1993) based on analyses of relearning in damaged networks. To this end, two simulations were carried out in the domain of reading via meaning to address a central issue in rehabilitation studies: what factors influence the degree of recovery and generalization observed in patients?

A general finding that emerged from the simulations is that the degree of recovery and generalization produced by retraining after damage depends strongly on the relative structure of the tasks performed by different parts of the system, and the extent to which the items selected for treatment approximate this structure. Specifically, the first stimulation found robust recovery and generalization from retraining following a lesion near semantic, but not following a lesion near orthography. In this way, the network results may help explain the variability in recovery observed in patients. The second simulation found, somewhat surprisingly, that retraining on less typical exemplars within a category produced greater generalization than did retraining on more typical exemplars. Overall, the results demonstrate that investigations of relearning after damage in connectionist networks can provide an account of the general nature of relearning and generalization in patients and can generate interesting hypotheses about the design of effective patient therapy.

References

Behrmann M (1987) The rites of righting writing: Homophone remediation in acquired dysgraphia. Cogn Neuropsychol 4:365–384

Behrmann M, Lieberthal T (1989) Category-specific treatment of a lexical semantic deficit: single case study of global aphasia. Brit J Commun Disorders 24:281–299

Coltheart M, Byng S (1989) A treatment for surface dyslexia. In: Seron X, Deloche G (eds) Cognitive approaches in neuropsychological rehabilitation. Hillsdale, NJ, Erlbaum, pp 159–174

Hillis AE (1993) The role of models of language processing in rehabilitation of language impairments. Aphasiology 7:5–26

Hinton GE, Plaut DC (1987) Using fast weights to deblur old memories. In: Proceedings of the 9th Annual Conference of the Cognitive Science Society. Hillsdale, NJ, Erlbaum 177–186

Hinton GE, Sejnowski TJ (1986) Learning and relearning in Boltzmann Machines. In: Rumelhart DE, McClelland JL, PDP Research Group (eds) Parallel distributed processing: Explorations in the microstructure of cognition. Volume 1: Foundations, Cambridge, MA, MIT Press, pp 282–317

Hinton GE, Shallice T (1991) Lesioning an attractor network: Investigations of acquired dyslexia. Psychol Rev 98:74–95

Kaas JH (1994) The reorganization of sensory and motor maps in adult mammals. In: Gazzangia MS, (ed) The cognitive neurosciences. Cambridge, MA, MIT Press, pp 51–71

Kertesz A (1985) Recovery and treatment. In: Heilman KM, Valenstein E (eds) Clinical neuropsychology. New York, Oxford University Press, pp 481–505

McClelland JL, Rumelhart DE, the PDP Research Group (eds) (1986) Parallel distributed processing: Explorations in the microstructure of cognition. Volume 2: Psychological and biological models. Cambridge, MA, MIT Press

Merzenich MM, Jenkins WM (1995) Cortical plasticity, learning and learning dysfunction. In: Julesz B, Kovacs I (eds) Maturational windows and adult cortical plasticity. Reading, MA, Addison-Wesley, 247–272

Merzenich MM, Jenkins WM, Johnson P, Schreiner C, Miller SL, Tallal P (1996) Temporal processing deficits of language-learning impaired children ameliorated by training. Science 271:77–81

Mozer MC, Behrmann M (1990) On the interaction of selective attention and lexical knowledge: A connectionist account of neglect dyslexia. J Cogn Neurosci 2:96–123

Plaut DC (1996) Relearning after damage in connectionist networks: Toward a theory of rehabilitation. Brain Language 52:25–82

Plaut DC (1997) Structure and function in the lexical system: Insights from distributed models of naming and lexical decision. Language Cogn Proc 12:767–808

Plaut DC, Shallice T (1993) Deep dyslexia: A case study of connectionist neuropsychology. Cogn Neuropsychol 10:377–500

Plaut DC, McClelland JL, Seidenberg MS, Patterson K (1996) Understanding normal and impaired word reading: Computational principles in quasi-regular domains. Psychol Rev 103:56–115

Posner MI, Keele SW (1968) On the genesis of abstract ideas. J Exper Psychol 77:353–363

Quinlan P (1991) Connectionism and psychology: A psychological perspective on new connectionist research. Chicago, University of Chicago Press

Robertson IE, Marshall JC (eds) (1993) Unilateral neglect: Clinical and experimental studies. Hillsdale NJ, Erlbaum.

Rosch E (1975) Cognitive representations of semantic categories. J Exper Psychol General 104:192–233

Rumelhart DE, Hinton GE, Williams RJ (1986a). Learning internal representations by error propagation. In: Rumelhart DE, McClelland JL, PDP Research Group (eds) Parallel distributed processing: explorations in the microstructure of cognition. Volume 1. Foundations. Cambridge, MA, MIT Press, pp 318–362

Rumelhart DE, McClelland JL, the PDP Research Group (eds) (1986b) Parallel distributed processing: explorations in the microstructure of cognition. Volume 1. Foundations. Cambridge, MA, MIT Press

Scott C, Byng S (1989) Computer assisted remediation of a homophone comprehension disorder in surface dyslexia. Aphasiology 3:301–320

Seidenberg MS, McClelland JL (1989) A distributed, developmental model of word recognition and naming. Psychol Rev 96:523–568

Tallal P, Miller S, Fitch RH (1993) Neurobiological basis of speech: a case for the preeminence of temporal processing. In: Tallal P, Galaburda AM, Llinas RR, von Euler E (eds) Temporal information processing in the nervous system: Special reference to dyslexia and dysphasia. New York, New York Academy of Sciences, pp 27–47

Tallal P, Miller SL, Bedi G, Byma G, Wang X, Nagaraja SS, Schreiner C, Jenkins WM, Merzenich MM (1996) Language comprehension in language-learning impaired children improved with acoustically modified speech. Science 271:81–84

Weekes B, Coltheart M (1996) Surface dyslexia and surface dysgraphia: Treatment studies and their theoretical implications. Cogn Neuropsychol 13:277–315

Some Neurological Principles Relevant to the Origins of – and the Cortical Plasticity-Based Remediation of – Developmental Language Impairments

M. M. Merzenich[1,3], P. Tallal[2,3], B. Peterson[3], S. Miller[3], and W. M. Jenkins[3,1]

Introduction

Recent advances in integrative neuroscience have led to an increasing under-standing of the brain plasticity mechanisms underlying perceptual, cognitive and motor skill learning in children and in adults (see Merzenich et al. 1991 a, b; Mer-zenich and Sameshima 1993; Merzenich and DeCharms 1996; Buonomano and Merzenich 1998, for reviews). These studies have shown how these natural neuro-logical processes can contribute to the often-powerful behavioral expressions of chronic neurological illnesses and disabilities (see Merzenich et al. 1991a, 1993, 1996a, b; Merzenich and Jenkins 1995; other chapters, this volume). They also point to practical strategies by which the same dynamic brain processes can be marshalled for neuroscience-guided, training-based amelioration of specific neu-rological impairments.

In this report, several features of cortical plasticity mechanisms underlying the normal development of human skills and abilities will be reviewed. Some of the ways in which these dynamic brain processes contribute to the expressions of human disabilities are outlined. With that background, an example of a neuroscience-guided training strategy applied for the remediation of develop-mental deficits in language learning is reviewed. Evidence is presented that neuroscience-based training can positively impact attentional, short-term mem-ory, receptive, cognitive and motor speech capabilities in children with language impairments. Finally, a hypothesis will be posited that addresses some aspects of the origins of the broad-ranging remedial effects of this intensive skills training.

Cortical Plasticity Mechanisms: An Overview

The detailed neuronal connections to and between cortical neurons are being continually remodeled by our experiences. The phenomenological nature of those changes have been documented in numerous experiments in which the

---

[1] Keck Center for Integrative Neurosciences University of California at San Francisco San Francisco CA 94143-0732

[2] Center for Molecular and Behavioral Neuroscience Rutgers University 197 University Avenue Newark NJ 07102

[3] Scientific Learning Corporation 1995 University Avenue Berkeley CA 94704-1074

J. Grafman / Y. Christen (Eds.)
Neuronal Plasticity:
Building a Bridge from the Laboratory to the Clinic
© Springer-Verlag Berlin Heidelberg New York 1999

cortex has been engaged to change by the introduction of new experiences or by the training of new perceptual, cognitive or motor skills (e. g., Jenkins et al. 1990, Recanzone et al. 1992a–c 1993; Wang et al. 1995; Nudo et al. 1996; see Merzenich and DeCharms 1996; Cruikshank and Weinberger 1996; Buonomano and Merzenich 1998 for reviews). In all such studies, cortical representations change during learning to create selective neuronal response representations of important new stimuli and behaviors.

Cortical networks specialize through learning to more selectively represent behaviorally important inputs and outputs. A main consequence of cortical plasticity mechanisms in learning is the amplification, in the specific neurological terms outlined below, of the representations of the inputs and outputs that are engaged by, or are effected in, the learned behavior. For a sensory guided movement behavior, for example, that specialization in distributed neuronal response representation commonly applies for manifold aspects and levels of sensory feedback information and of movement planning and control. As an animal acquires a sensory guided hand movement skill, like the successful retrieval of a small object under constrained mechanical conditions, changes can be recorded in sensory representations in a cortical area that represents predominantly cutaneous tactile inputs in cortical area 3b (e.g., Jenkins et al. 1990; Xerri et al., submitted for publication; also see Pascual-Leone et al. 1995; Elbert et al. 1995), in a second sensory representation of cutaneous afferents in cortical area 1 (Merzenich et al. 1991b, and in a sensory representation of muscle spindle and other deep afferents in cortical area 3a (Recanzone et al. 1992b). Other cutaneous, proprioceptive and kinesthetic sensory zones (e.g., cortical areas 2, 5, 7, SII and insular cortical areas) are also directly engaged to change by any demanding (closely attended), manual sensory-guided behavior.

Remodeling by manual sensory-guided movement behaviors have also been documented in motor planning and motor output zones in the frontal lobe. Large scale behavior-specific response remodeling has been recorded, for example, in movement representations controlling input sequences and sensory feedback-movement associations in multiple divisions of cortical area 6 (e.g., see Mitz et al. 1991; Aizawa et al. 1991), in several functionally different motor planning areas in the supplementary motor cortex (e.g., Tanji et al. 1996), and in the pyramidal tract output zones in cortical area 4 (e.g., Nudo et al. 1995; Pascual-Leone et al. 1995; Karni et al. 1995).

All of these 15–20 or more cortical areas are directly engaged by behaviorally important inputs and actions in this sort of sensory-guided manual skills behavior *to change.* In every directly engaged cortical zone, distributed neuronal response representations of inputs, action plans and actions progressively evolve in sensori-motor skill learning. It is reasonable to hypothesize that the sum of these distributed, multiple-field changes constitutes *the* cortical part of the learning of the skill (Merzenich et al. 1991a, b; Merzenich and Jenkins 1994; Merzenich and Sameshima 1993; Merzenich and DeCharms 1996).

The induction of at least much of this large scale change in distributed neuronal responses in the cortex follows Hebbian learning principles. One basis of this learning-induced "selection" of behaviorally important inputs is Hebbian plasticity (Hebb 1949). Hebb's great principle is that inputs that excite neurons nearly simultaneously in time change their effectivenesses together. By that synaptic plasticity principle, within the limitations of the anatomical spreads of inputs within any cortical field (Merzenich et al. 1984, 1991 a, b), the representations of any part of a behaviorally important input will be co-selected – integrated – to create complex stimulus- and action event-specific neuronal responses in the cortex.

Hebbian mechanisms also operate to increase the connection strengths between nearly simultaneously excited neurons *within* cortical networks, to create what Hebb (1949) termed "cortical cell assemblies." Connections that are active nearly simultaneously in time *including complex interconnections between cortical neurons* all strengthen together. By such network changes in positive coupling between nearly simultaneously excited neurons, stimulus and action event-specific "cell assemblies" (Hebb 1949) or "neuronal groups" (Edelman and Finkel 1984) are formed and grow in neuronal memberships and strengths progressively as training progresses.

With the resultant increasingly stronger positive coupling and as the memberships of cooperative neuronal groups grow, neurons "share" their selectively driven inputs with increasing selectivity and fidelity, and the individual neuronal elements of a growing cell assembly respond more and more synchronously in time (Fig. 1; Merzenich et al. 1991 a, b; Merzenich and Sameshima 1993; Wang et al. 1995; Merzenich and DeCharms 1996).

It should be noted that the actual governing cellular and molecular mechanisms are complex, and that there are several aspects of cortical change phenomenology in learning that are not Hebbian in origin. Several classes of negative change in synaptic effectivenesses (several mechanisms generating "long-term

1. With progressive training:
   a) cortical sectors representing behaviorally important inputs and actions grow;
   b) representational remodeling is recorded in *many* different, functionally specific cortical areas;
   c) representation changes are powerfully modulated as a function of behavioral state;
   d) connections to and between neuron members of growing neuronal cell assemblies progressively strengthen, following an input-coincidence-based (Hebbian) synaptic strengthening rule;
   e) increasingly more strongly coupled neurons in these cooperatively coupled assemblies represent behaviorally important input and action events with progressively increasing fidelity, by progressively stronger distributed neuronal response coordination;
   f) distributed neuronal response coordination is equatable with neuronal representational power (salience) in the cerebral cortex.
2. For multiple-component "complex stimuli" (i. e., for most real behavioral inputs and actions):
   a) inputs are processed by the cortex in integrated, variable-length time chunks;
   b) with appropriate complex stimulus reception training, as cell assemblies forms they come to represent successive inputs more discretely in time because the integrative time constants governing segmentation of input and action event streams are plastic.

Fig. 1. Some Principles Underlying the Creation of Distributed Neuronal Response Representations of Stimulus and Action Events in the Cerebral Cortex During Skill Learning

depression") are recorded in parallel with positive Hebbian changes ("long-term facilitation"; see Merzenich and DeCharms 1996; Buonomano and Merzenich 1998, for review). Mechanisms of plasticity of connections to and from inhibitory neurons in cortical circuits follow different change rules. At the same time, phenomenologically, Hebbian models provide a reasonable (if incomplete) description of the distributed changes recorded in learning-induced plasticity studies.

The cortex samples successive inputs in time "chunks." Cortical mechanisms are primarily concerned with recording rapidly changing or rapidly successive input and action events by the processing of information in narrow slices in time (see Merzenich et al. 1993; Merzenich and Jenkins 1995). Any substantially new incident input event "resets" cortical dynamics, and re-initiates input sampling or "recovery time" for taking another sample. The "reset" is a function of the content of both the immediately preceding input event and the "resetting" event. These dynamics apply primarily for cortical processing "channels;" rapid successive sampling can always occur across separate, parallel channels.

Note that if the integrative time constants of the cortical processing machinery that segments the incoming stream of information are too short, no adequately integrated samples of real-world inputs or activities could be obtained. If they are too long, the cortex will inadequately representationally devolve rapidly changing or rapidly successive features of serial or complexly dynamically changing inputs.

The integrative time constants of cortical processing machinery that control this input event sampling are themselves plastic. The basic effectiveness with which the cortex can derive accurate, recognizable samples across time is subject to powerful plasticity effects (see Merzenich et al. 1993; Merzenich and Jenkins 1995). In psychophysical studies in humans, training can result in the accurate reception of successive inputs delivered to the same cortical processing channels at progressively higher speeds. For example, the identification of a brief visual probe is initially powerfully interfered with by the presentation of a following visual masker unless it is separated by a relatively long intervening time period. However, the required „cortical processing time" or successive-sample "recovery time" separating the first stimulus (the probe) from the second (the masker) can be dramatically shortened by intensive training (Karni and Sagi 1991, 1993; Ahissar and Hochstein 1993). Similar effects have been recorded in monkeys and in humans in somesthesis and in audition (Merzenich and Jenkins 1995; Merzenich et al. 1996a, b). *With intensive practice, a cortical sector can be "trained" to shorten (or lengthen) its cortical integration ("processing times"), i. e., to more rapidly and efficiently segment the representations of rapidly changing or successive events* and to thereby represent, by distributed neuronal discharges, ongoing or rapidly successive acoustic features with progressively greater time resolution and precision (Fig. 1).

Some of these learning-induced changes generalize in cortical receiving areas to only a limited extent; other learning-induced plasticity changes generalize broadly. When a monkey is trained at a spectral (spatial) discrimination task, changes induced in the cortical network are relatively local, primarily involving the directly engaged skin surfaces and cortical zone of representation of immediately surrounding skin (e.g., Jenkins et al. 1990; Recanzone et al. 1993; Wang et al. 1995; also see Pavlov 1927; Cruikshank and Weinberger 1996). When a monkey or human is trained to detect time-varying features of response for spatially located stimuli, distributed neuronal changes induced in the network are again relatively local in the primary receiving areas of the cortex (A-I or S-I; e.g., Recanzone et al. 1992 a–c, 1993; Wang et al. 1995; Ahissar and Hochstein 1993). For example, if a spectrotemporal signal recognition/discrimination training task involves stimulation at a given point on the skin of the hand, stimulation at a given sound frequency or stimulation at a given point in the visual field, cortical network changes and/or behavioral generalization have been recorded 1) for the representations of adjacent digits in area 3 b (S-I) but not more broadly across the hand surfaces, 2) over the zone of representation in A-I of a significant part of a sound frequency octave but with no evident effect on zones of more distant representation of still higher or still lower sound frequencies, and 3) restricted within a limited visual field sector.

By contrast, if a subject is trained in a purely temporal discrimination task (e.g., to discriminate differences in the time separations between stimuli around a particular base time interval, e.g., 100 ms), training-driven performance improvements generalize broadly, e.g., from a trained tonal frequency to across the zone of representation of at least several octaves of sound frequency, widely from a trained point on the skin out across the ipsilateral and contralateral body surface, and, indeed, to some extent across modality (see Wright et al. 1997 b; Nagarajan et al. 1998 a). At the same time, training-based improvements in discriminative abilities in a purely temporal task generalize far less strongly to other base time intervals.

The local cortical network change mechanisms underlying local spectrotemporal feature learning are understood on a first level, albeit very incompletely when considered in mechanistic terms (e.g., Merzenich and DeCharms 1996; Cruikshank and Weinberger 1996; Gilbert et al. 1996; Buonomano and Merzenich 1998). The source(s) of, and the mechanisms underlying, the neurological plasticity accounting for learning and generalization in purely temporal tasks is less well understood (see Ivry 1996; Nagarajan et al. 1998 a). Whatever their specific neurological origins, generalization of training effects is a primary consideration for the creation of any therapeutically useful remediative training regime.

Learning-induced changes are modulated as a function of the behavioral state of the animal. In the primary auditory, somatosensory and motor cortices that have been the main focus of our electrophysiological studies of learning-induced brain plasticity, long-term changes have been induced by closely attended behaviors, but not when equivalent schedules of input or action events occur without the monkey attending to them (Recanzone et al. 1992 a–c, 1993, Nudo et al. 1995; also see Ahissar et al. 1992; Kilgard and Merzenich 1998), or when they can be received or performed under relatively automatic (unattended) behavioral performance conditions (Nudo et al. 1996; see Frackowiak, this volume). That does not mean, of course, that behaviorally unattended inputs are not received and represented transiently in some form in even these primary cortical receiving areas. Neither does it mean that other cortical fields are not induced to change in a more substantial long-term manner by implicit inputs (see Frackowiak, this volume).

It is important to emphasize that this is not simply an "on-off" switch for permitting or disallowing enduring plasticity changes to occur. The enabling feedback is *graded* to reflect the importance of the input to the animal or man. Behavioral attention and high motivation for correct performance are obviously critically important for efficiently driving positive distributed neuronal representational change in any practical rehabilitative setting.

The scale of learning-induced cortical change can be massive. When a monkey is trained in a simple sensory-guided motor behavior that it masters in a few hundred or a few thousand training trials, as noted above, neuronal responses are enduringly positively modified during the acquisition phase of the behavior in many sensory and motor cortical areas. *The selective responses of millions or tens of millions of neurons are altered by such training.*

Similarly, when a human subject is trained in a simple sensory motor training task refined through only a few hours of intense practice, very significant representational changes are relatively quickly generated in the cortex for both the "sensory" and "motor" aspects of the task, just as in monkeys. Again, changes induced in such intensive training behaviors must involve alterations of neuronal response selectivity in millions or tens of millions of neurons, and potentially arises from changes in numbers and/or strengths of hundreds of millions or possibly billions of synaptic contacts.

When one considers the representational changes that must underlie the creation of progressively differentiated and elaborated movement repertoires, or that underlie speech reception and language usage across the course of child development – involving, in both cases, literally *millions* of attended "practice" input and action events – *local synaptic connectivity changes and distributed cortical neuronal response changes are occurring on a massive scale.*

These learning-induced changes are not strictly governed by a "critical period"; fundamental mechanisms supporting use-dependent plasticity are in place throughout life. While critical period phenomena clearly apply for early developmental processes, there is of course no critical period disabling of these dynamic learning processes. The continuous, adaptive changes that mark "child development" are almost certainly primarily attributed to these functional self-creation processes. They operate continuously and seamlessly across childhood, indeed, across life. Limitations in the brain's capacity to facilely generate representations of complex behaviors – for example, a second language – have often been interpreted to mean that the brain's fundamental change processes are attenuated in some physical and irreversible way. To the contrary, we have hypothesized that they are probably primarily attributed to learning success, i.e., to the establishment of progressively more powerful representations of the knowledge and expectation structures that guide use of the primary language, and not to any down-regulation of the learning machinery *per se* (see Merzenich and Jenkins 1994; Merzenich and DeCharms 1996; Merzenich et al. 1996 a)

In learning, the brain creates progressively more salient representations of stimulus and action events. As stimulus-specific cell assemblies form in learning by an increase in the strengths of coupling between the neurons that comprise them, they come to represent inputs in an increasingly robust and more salient form. In early learning especially, cell assembly neuronal memberships grow rapidly. Perhaps more importantly, the distributed coordination in the responses of coupled neurons grows substantially, as positive coupling between neuronal constituents increases. Distributed coordination of discharges provides an amplification of the *power* of distributed neuronal response representations of the behaviorally important stimulus or action event, because all destinations of the outputs of cortical neurons are, similarly, input coincidence-dependent neuronal integrators operating with relatively short integrative time constants.

Positive changes in representational salience should reflect memory and cognitive operations. After learning, in a given cortical area tens or hundreds of thousands or potentially millions of neurons might be participating in the distributed representation of a behaviorally important stimulus or action. In any given cortical area, there has been a progressive increase in the distributed coordination of the discharges of these neurons. Now, in each zone, many thousands or hundreds of thousands or millions of neurons are acting temporally in concert, while before, their activities were substantially more temporally dispersed (Merzenich and Sameshima 1993; Merzenich and DeCharms 1996). In their output projections, these *intensively practice-driven neuronal networks in A1 and SI deliver a proportionally more highly salient message to the downstream cortical and extracortical areas that they feed.* At all of those destinations, involved in other aspects of memory and cognition and action control, neurons receiving these outputs downstream are temporal integrators operating with relatively short time constants. Hypothetically, the more salient the signal delivered to them, the more reliable their ("memory", "cognitive") operations.

Summary of learning phenomenology. Some of these most basic aspects of cortical phenomenological change induced by skill learning are summarized schematically in Figure 1. To summarize in perceptual, cognitive or motor skill learning, coincident input "co-selection" mechanisms (Hebbian plasticity) operate to create selective representations of behaviorally important inputs and actions. For any trained (attended) behavior, local changes in input effectivenesses are induced in many cortical areas, with each contributing in an area-specific way to progressive skill learning. Through coincident input "co-selection" mechanisms, input- and output-specific cooperative neuronal groups (cell assemblies) are formed by positive changes in coupling (synaptic) strengths for inputs delivered to cortical networks, and that interconnect cortical neurons. Distributed discharge coordination in input or action event representation grows as neuronal cell assembly memberships and these network coupling strengths grow. That results, effectively, in an increase in the power of the distributed neuronal representation of these behaviorally important input and action event activities. This amplified and more salient neuronal representation more power-fully engages its extrinsic output projection targets.

These processes operate on input samples that are integrated over limited time epochs. Any substantially new input event resets the cortical dynamics, and with a brief delay, re-initiates the derivation of a new, distributed neuronal response sample. The integrative time constants and the processes that govern the time course of these sequenced excitatory-inhibitory effects (sampling rates) are themselves subject to powerful learning effects. The cortex can be trained to derive samples across relatively long or relatively short integration (input event segmentation) times. *Training-based improvements of processing speed can greatly improve the distributed, coordinated neuronal response representations (the salience) of the fine spectrotemporal details of incoming input streams.*

Contributions of Cortical Plasticity Processes to Human Disability; The Example of Developmental Language Impairments

Children with language-based learning impairment (LLI) apparently process rapidly changing and rapidly successive acoustic inputs in a different manner than do "normal" children. These children receive poorly spectrotemporally resolved speech, which is manifested by poor phonological reception (see Bishop 1992; Tallal et al. 1993; Merzenich et al. 1998 for reviews). They make many errors in identifying the sound parts of speech, have difficulties in parsing words into their sound parts, have significantly impaired memories for aurally received speech, have limited language comprehension abilities, and have many deficits in the cognitive operations that apply in language usage.

Psychophysical studies indicate that children with language learning impair-ments are processing acoustic inputs with the cortical machinery operating with relatively long integrative time constants. This abnormal signal "chunking" is manifested by abnormal detection masking and recognition masking abilities (e. g., see Tallal and Piercy 1973, 1974, 1975; Tallal et al. 1993; Farmer and Klein 1995; Wright et al. 1997a). Hypothetically, because of the abnormally long inte-gration times for their cortical processing machinery, these children do not seg-ment the rapidly changing or successive parts of complex acoustic inputs – e.g., the sound parts of words – in the normal way. When a brief, weak probe tone is presented in front of a masker, children with LLI cannot detect it (apparently because of abnormally prolonged signal integration in the cortex) unless it is ele-vated in intensity by more than two orders of magnitude in comparison with a normal child, or unless it is separated in time from the masker by several hun-dred milliseconds, under conditions in which a normal child can detect it when it immediately precedes the same masker (Wright et al. 1997a). Similarly, the brief initial parts of a consonant-vowel (CV) stimulus can be masked by an immediately following vowel (Tallal and Piercy 1973, 1975). Language-impaired children not uncommonly confuse a synthetic CV like a /ba/ or /da/ with the pre-sentation of the V alone, apparently because abnormally prolonged integration results in the degradation of brief consonant transitions (see Tallal and Piercy 1974, 1995; de Wierdt 1989). The frequency-transition events that signal consonant identity can be rendered audible in these children – can be unmasked – by increasing its duration, by differentially increasing its intensity, by exaggerating its spectral differences from confusable CVs, or by separating it in time from the following vowel.

Even when the initial sound in a two-stimulus sound sequence is clearly heard separately in a language impaired child, their abnormal signal integration results in strong "recognition masking" marked by a degradation of the impaired subject's ability to make fine spectrotemporal distinctions about it.

This fundamental difference in signal integration/segmentation in acoustic signal reception in children with LLI emerges in the first year of life (Benasich and Tallal 1996). The ability of a six-month-old infant to identify rapidly successive stimuli has normally evolved to achieve selective, high-speed acoustic processing capabilities. By contrast, in many of the children born to families with a history of LLI, this high-speed processing does not develop in the normal way. For them, powerful recognition masking effects enduringly impair their ability to accurately identify and sequence rapidly successive sounds within a given sound processing channel unless they are separated in time by several hundred milliseconds. Not surprisingly, these children are later almost invariably identified as "language delayed" as 2- and 3-year-olds (Benasich and Tallal 1996).

We have hypothesized that the heavy, continued use of relatively long integration times is an expected consequence of a system that has to make distinctions about poorly spectrotemporally resolved inputs in the first year of life (Merzenich et al. 1993; Merzenich and Jenkins 1995). For example, it would predictably characterize cortical networks that were making decisions about inputs delivered or processed under relatively high signal-to-noise conditions, where longer integration times would be required to make reliably rewardable distinctions. This problem could arise because of poor sound resolution by the ear, e.g., arising from an early chronic conductive hearing loss. It could also arise from any central nervous system deficit(s) that degraded the signal-to-noise treatment of neurological representations of complex acoustic inputs.

Whatever its specific origin (and there are almost certainly multiple specific origins), many young school-age children (probably well over 10%) appear to process acoustic information including speech in this alternative way. Their abnormally "slow" processing originates in and has apparently been deeply embedded by brain plasticity mechanisms that operate to create speech reception and language representations across early childhood.

An Example of a Cortical Plasticity-Based Remediation Strategy: A Training Program Designed to Ameliorate the Acoustic Signal (Speech) Processing Deficits of Language Learning-Impaired Children

Basis of Design of the Training Program. We know that, from very early childhood, children with LLIs are processing complex acoustic inputs including speech with an abnormally prolonged temporal integration of incoming acoustic signals. This signal reception deficit hypothetically underlies their degraded ability to resolve the fine spectrotemporal details of speech and has manifestly resulted in the establishment, through the course of child development, of an abnormal form of distributed neurological representation of aurally-received speech.

From studies of the plasticity of the cortical neuronal representations of complex signals, we know that the time constants governing the refined processing of rapidly changing or rapidly successive input events can be altered by intensive practice. At any point in life, *the brain can be* <u>*trained*</u> *to make more accurate distinctions about rapidly changing or rapidly successive inputs.* It should be possible to shorten the time constants governing the integration/segmentation processing times in these children with LLI with appropriate training.

From studies of cortical plasticity and from experimental psychology and psychophysics, we know that any substantially complete "correction" of this deficit in a language-impaired child will necessarily require intensive training. From studies of training generalization effects, we also know that extensive training across the spectral and temporal dimensions that apply for accurate complex signal (speech) reception shall be required to address all of the spectrotemporal contextual conditions that apply for speech reception and language usage.

From many studies in cognitive neuroscience documenting the neuromodulatory processes that "gate" the learning machinery, we know that the rapid induction of appropriate representational changes require closely attended and highly rewarded behaviors.

On the basis of these and other related premises, a set of novel, computer-based, cognitive neuroscience-guided training tools, which we called "Fast For-Word™" were created. These psychophysical training exercises have been described in detail elsewhere (see Merzenich et al. 1998; www.FastForWord.com; also see Nagarajan et al. 1998b). Exercises were "disguised" as computer "games" in which all important decisions in the game required the child with LLI to make fine listening distinctions. In training, children with LLIs were maintained under close behavioral control by the exercise, and received a rich array of feedback information and rewards. Each of seven exercises was self-adjusted so that the children initiated training under conditions in which they could always make accurate aural percept distinctions. Using an adaptive training procedure format, stimuli were adjusted continuously in difficulty toward normalcy as the child progressed in each listening skill. Each exercise began with the use of acoustically modified inputs that were differentially spectrally and temporally exaggerated to facilitate the child's ability to make correct behavioral distinctions. The end point of each game, achieved after several hours of adaptive training at each of these seven hierarchically organized behavioral tasks, was a normal ability to resolve the fine spectrotemporal features of complex acoustic inputs or speech, and/or to receive and operate with speech inputs with a normal age-appropriate competence.

What was achieved with the application of this neural plasticity-based training program in a population of children with language impairments? With this training approach, we have now demonstrated that we can rapidly drive very large improvements in the fundamental acoustic reception, speech reception and language abilities of children with LLI (see Tallal et al. 1996; Merzenich et al. 1996c; Merzenich et al. 1998 for further details). Some results of a trial conducted with more than 500 of these children are shown in Figures 2–5. Results from an elementary speech reception test, the Goldman-Fristoe-Woodcock Test of Auditory Discrimination (1970), are shown in Figures 2 and 3. In this simple acoustic speech reception test, the child is presented with a word along with pictures that represent that target word and aurally confusable words. High performance on the task requires that the child accurately hear the sound parts of the target word, and that he/she understands and can signal (by picture pointing) its meaning.

Children with LLIs trained in this trial were impaired at this fundamental task (Figs. 2 and 3, upper panels). After an average total of about 50 hours of training at the seven *Fast Forword* exercises, the performance of boys with language impairments advanced by an average of more than one Standard Deviation compared to the normal distribution for the test administered in a quiet background (Fig. 2), and by about two Standard Deviations compared to the normal distribution when the test was administered with background noise (simulating the noise level of an average American classroom). Essentially equivalent changes were recorded in trained LLI girls. Very positive changes were recorded, on the average, in children with LLI of all ages in this trained population.

Note that significant numbers of individual, trained children (Figs. 2 and 3, lower panels) improved in this fundamental speech reception ability by two, three, four or five Standard Deviations compared to the American population. These represent very large advances in their fundamental speech reception competencies.

These results are also important, of course, because this is precisely the fundamental skill – the more accurate reception of the rapidly changing, fine spectrotemporal acoustic features of speech – that this neuroscience-guided training program was designed to improve. After training, from an initially "impaired" (< 1 SD below the American population mean) level of performance, *the average trained LLI boy or girl approached or exceeded the normal median performance for children of the same age in the American population* (Figs. 2 and 3, upper panels). It is concluded that this very fundamental aspect of their speech reception abilities has been substantially remediated in this population by this approximately 50 hours of intensive listening skills training.

Interestingly, benefits extended well past aural speech reception abilities to improvements in more general language abilities. Improvements in overall lan-

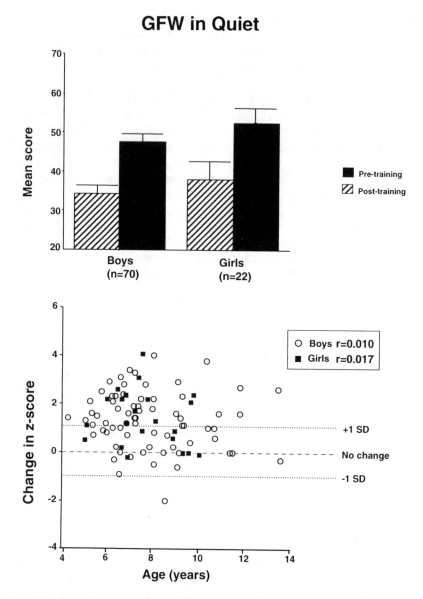

Fig. 2. Upper panel: Results obtained on the Goldman-Fristoe-Woodcock (GFW) Test of Auditory Dis-crimination (1970) with the test administered in a Quiet (low-noise) background before (cross-hatched) and after (black) training. Thin bars are standard errors. Results were highly significant (p<o.ooo1). Data from all 92 of the children with LLI in which this specific test was administered in a large (> 500 child) field trial of *Fast ForWord* are shown. **Lower panel:** Pre- vs post-training score changes, plotted as z-score (standard deviation) changes, for all 92 individual *Fast ForWord*-trained children. Score change data are plotted as a function of the age of the child with LLLI

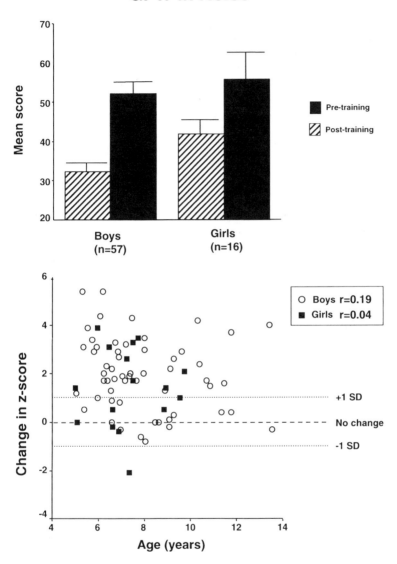

Fig. 3. Upper panel: Results obtained on the Goldman-Fristoe-Woodcock (GFW) Test of Auditory Discrimination (1970) with the test administered with a Noise background simulating the noise in a typical Americal classroom before (cross-hatched) and after (black) training. Thin bars are standard errors. Results were highly significant for both girls and boys (p<o.oool). Data from all 73 children with LLI in which this specific test was administered are shown. **Lower panel:** Pre- vs post-training score changes, plotted as z-score (standard deviation) changes, for all 73 individual trained children. Score change data are plotted as a function of the child's age

guage ability are commonly indexed in this population by a "language battery" that, like an IQ test, gauges the specific language competence of a child with reference to the American population. Results from 167 children given different age-appropriate versions of two standard American language batteries (Test of Language Development, TOLD; Clinical Evaluation of Language Function, CELF) are shown by example in Figures 4 and 5. Again, very substantial average advances toward the normal mean were recorded for measures of the overall "language quotient" (LQ) for both boys and girls, with performance abilities approaching the normal mean (an LQ of 100) after training (Figs. 4 and 5, upper panels). This represents an average advance of the effective language age of the child of about 1.5 years, achieved with about 50 hours of training over a five-week training period.

The training in *Fast ForWord* involved only listening skills training. It is important to note, then, that its impacts extended immediately beyond listening skills improvement to speaking skills improvements (Fig. 5). Significant improvements were also recorded on measures of memory and cognitive abilities in this population. In fact, 57 different standardized language-related tests assessing numerous facets of speech reception, syntax, grammar and speech usage requiring cognitive, memory and production skills were administered to more than 10 children/test in this large trial. Statistically significant score advances were recorded in 55 of these tests; pre-vs post-training score changes were significant at the $p > 0.0001$ level in 47 of them.

Some Implications. The sweeping, positive changes recorded in children trained with this neuroscience-guided program are consistent with our understanding of how skill learning drives changes in the distributed neuronal response representations of stimulus and action events in the cerebral cortex. If training like that applied in these exercises in children with LLI were to be applied in a rat or monkey model, we know that more refined stimulus-specific and temporally discrete representations of rapidly succesive and rapidly changing inputs would emerge. With the resultant higher-speed processing, cortical integration (processing) times could be shortened from the many tens to several hundred millisecond-long times that apply for the majority of children with LLI in this large sample, down to the normal few tens of milliseconds that apply for high-fidelity, fast processing in normal complex acoustic signal (speech) reception.

In a monkey, those changes would unequivocally generate more temporally coordinated distributed cortical neuronal network responses. The sound parts of words would thereby be represented with greater representational power. We hypothesize that such increased neurological salience underlies the sweeping, positive impacts on higher-level speech and language tests, as well as the often-striking improvements in expressive abilities recorded in this trial.

Consider a simple analogy. Given blurred vision, performance measures on a wide variety of perceptual, cognitive and visually guided performance tasks will be poor. Performance improvements will be recorded at all of those tasks by clarifying vision with corrective lenses. In an analogous way, these training

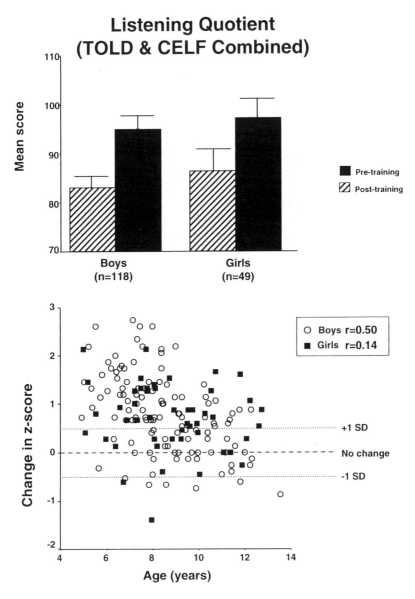

Fig. 4. Upper panel: Listening (Receptive Language) Quotients obtained for the TOLD (Newcomer and Hammill 1988) and CELF (Semel et al. 1995) language batteries (see text) combined, for the test administered before (cross-hatched) and after (black) training. Thin bars are standard errors. Results were highly significant (p<0.0001). Combination of TOLD and CELF battery data was statistically appropriate. Data from 167 children with LLI in which these specific test batteries were administered in a large field trial of *Fast ForWord* are shown. **Lower panel:** Pre- vs post-training score changes, plotted as z-score (standard deviation) changes, for all 167 individual *Fast ForWord*-trained children, plotted as a function of the child's age

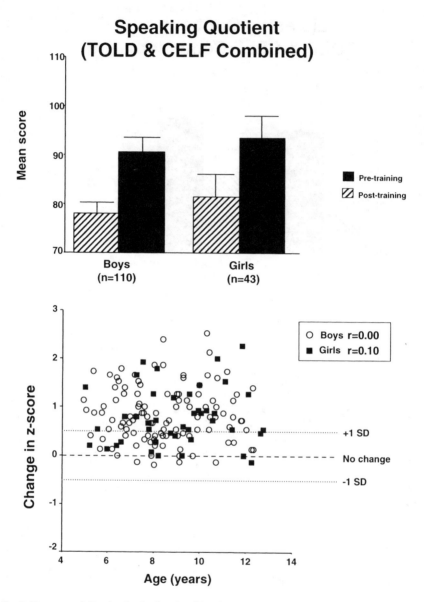

Fig. 5. Upper panel: Results obtained on Speaking (Expressive Language) Quotients for the TOLD and CELF language batteries (see text) combined, for the test administered before (cross-hatched) and after (black) training. Thin bars are standard erros. Results (post-training vs pre-training scores) were highly significant (p<0.001). Combination of TOLD and CELF battery data was statistically appropriate. Data from 167 children with LLI in which this specific test was administered in a large field trial of *Fast ForWord* are shown. **Lower panel:** Before- vs after-training score changes, plotted as z-score (standard deviation) changes, for all 167 individual *Fast ForWord*-trained children. Score change data are plotted as a function of the LLI child's age

results, coupled with our growing understanding of brain plasticity/learning phenomenology, indicate that this intensive listening clarifies the distributed neuronal representations of the acoustic fine-structure of speech. That should – and apparently does – have correspondingly broad-ranging, positive neurobehavioral consequences.

References

Ahissar M, Hochstein S (1993) Attentional control of early perceptual learning. Proc Natl Acad Sci USA 90:5718–5722

Ahissar E, Vaadia E, Ahissar M, Bergman H, Arieli A, Abeles M (1992) Dependence of cortical plasticity on correlated activity of single neurons and on behavioral context. Science 257:1412–1415

Aizawa H, Inase M, Mushiake H, Shima K, Tanji J (1991) Reorganization of activity in the supplementary motor area associated with motor learning and functional recovery. Exp Brain Res 84:758–761

Benasich AA, Tallal P (1996) Auditory temporal processing thresholds, habituation, and recognition memory over the first year. Infant behav Develop 19:339–357

Bishop DV (1992) The underlying nature of specific language impairment. J Child Psychol Psychiat Allied Discip 33:3–66

Buonomano D, Merzenich MM (1998) Cortical plasticity: from synapses to maps. Ann Rev Neurosci 21:149–186

Cruikshank SJ, Weinberger NM (1996) Evidence for the Hebbian hypothesis in experience-dependent physiological plasticity of neocortex: a critical review. Brain Res Rev 22:191–228

Deiber MP, Wise SP, Honda M, Catalan MJ, Grafman J, Hallett M (1997) Frontal and parietal networks for conditional motor learning: a positron emission tomography study. J Neurophysiol 78:977–991

Edelman GM, Finkel LM (1984) Neuronal group selection in the cerebral cortex. In: Edelman GM, Gall WE, Cohen WM (eds.) Dynamic aspects of neocortical function. Wiley, New York, pp 653–695

Elbert T, Pantev C, Wienbruch C, Rockstroh B, Taub E (1995) Increased cortical representation of the fingers of the left hand in string players. Science 270:305–307

Farmer ME, Klein R (1995) The evidence for a temporal processing deficit linked to dyslexia: a review. Psychonom Bull Rev 2:460–493

Gilbert CD, Das A, Ito M, Kapadia MK, Westheimer G (1996) Cortical dynamics and visual perception. Cold Spring Harb Symp Quant Biol 61:105–113

Goldman R, Fristoe M, Woodcock RW (1970) Goldman-Fristoe-Woodcock test and auditory discrimination. Circle Pines, MN, American Guidance Service Inc

Hebb DB (1949) Organization of behavior. Wiley, New York

Ivry RB (1996) The representation of temporal information in perception and motor control. Curr Opin Neurobiol 6:851–857

Jenkins WM, Merzenich MM, Ochs MT, Allard T, Guic RI (1990) Functional reorganization of primary somatosensory cortex in adult owl monkeys after behaviorally controlled tactile stimulation. J Neurophysiol 63:82–104

Karni A, Sagi D (1991). Where practice makes perfect in texture discrimination: evidence for primary visual cortical plasticity. Proc Natl Acad Sci USA 88:4966–4970

Karni A, Sagi D (1993) The time course of learning a visual skill. Nature 365:250–252

Karni A, Meyer G, Jezzard P, Adams MM, Turner R, Ungerleider LG (1995) Functional MRI evidence for adult motor cortex plasticity during motor skill learning. Nature 377:155–158

Kilgard M, Merzenich MM (1998) Cortical map reorganization enabled by nucleus basalis activity. Science 279:1714–1718

Merzenich MM, DeCharms RC (1996) Neural representations, experience and change. In: Llinas R, Churchland P (eds.) The mind-brain continuum, MIT Press, Boston, pp. 61–81

Merzenich MM, Jenkins WM (1994) Cortical representation of learned behaviors. In: Anderson, P, Hvalby O, Paulsen O, Hökfelt B (eds.) Memory concepts. Amsterdam, Elsevier, pp 437–450

Merzenich MM, Jenkins WM (1995) Cortical plasticity, learning and learning dysfunction. In: Julesz B, Kovacs I (eds.) Maturational windows and adult cortical plasticity. Addison-Wesley, pp 247–272

Merzenich MM, Sameshima K (1993) Cortical plasticity and memory. Curr Opin Neurobiol 3:187–196

Merzenich MM, Jenkins WM, Middlebrooks JC (1984) Observations and hypotheses on special organization features of the central auditory nervous system. In: Edelman GM, Gall WE, Cowan WM (eds.) Dynamic aspects of neocortical function. Wiley, New York

Merzenich MM, Grajski KA, Jenkins WM, Recanzone GH, Peterson B (1991a) Functional cortical plasticity. Cortical network origins of representational changes. Cold Spring Harb Symp Quant Biol 55:873–887, 1991

Merzenich MM, Recanzone GH, Jenkins WM (1991b) How the brain functionally rewires itself. In: Arbib M, Robinson JA (eds.) Natural and artificial parallel computations. MIT Press, New York

Merzenich MM, Schreiner C, Jenkins W, Wang X (1993) Neural mechanisms underlying temporal integration, segmentation, and input sequence representation: Some implications for the origin of learning disabilities. Ann NY Acad Sci 682:1–22

Merzenich MM, Spengler F, Byl N, Wang X, Jenkins W (1996a) Representational plasticity underlying learning; contributions to the origins and expressions of neurobehavioral disabilities. In: Ono T, McNaughton BL, Molotchnikoff S, Rolls ET, Nishijo H (eds.) Perception, memory and emotion: frontiers in neuroscience. Pergamon, Cambridge

Merzenich MM, Wright B, Jenkins W, Xerri C, Byl N, Miller S, Tallal P (1996b) Cortical plasticity underlying perceptual, motor and cognitive skill development: implications for neurorehabilitation. Cold Spring Harb Symp Quant Biol 61:1–9

Merzenich M, Jenkins W, Johnston P, Schreiner C, Miller SL, Tallal P (1996c) Temporal processing deficits of language-learning impaired children ameliorated by training. Science 271:77–80

Merzenich MM, Miller S, Jenkins W, Sauners G, Protopapas A, Peterson B, Tallal P (1998) Amelioration of the acoustic reception and speech reception deficits underlying language-based learning impairments. In: Euler CV (ed.) Basic neural mechanisms in cognition and language. Amsterdam, Elsevier

Mitz AR, Godschalk M, Wise SP (1991) Learning-dependent neuronal activity in the premotor cortex: activity during the acquisition of conditional motor associations. J Neurosci 11:1855–1872

Nagarajan SS, Wang X, Merzenich MM, Schreiner CE, Johnston PA, Jenkins WM, Miller SL and Tallal P (1998) Speech modification algorithms used for training language-learning impaired children (LLIs) IEEE Trans Rehab Eng, in press

Nagarajan SS, Blake DT, Wright BA, Byl N, Merzenich MM (1998b) Practice-related improvements in somatosensory interval discrimination is temporally specific but generalizes across skin location, hemisphere and modality. J Neurosci, 18:1559–1570

Newcomer PL, Hammill DD (1988) Test of language development primary, Second Edition. Austin, TX, Pro-Ed

Nudo RJ, Milliken GW, Jenkins WM, Merzenich MM (1996) Use-dependent alterations of movement representations in primary motor cortex of adult squirrel monkeys. J Neurosci 16:785–807

Pascual-Leone A, Wassermann EM, Sadato N, Hallett M (1995) The role of reading activity on the modulation of motor cortical outputs to the reading hand in Braille readers. Ann Neurol 38:910–915

Pascual-Leone A, Wassermann EM, Grafman J, Hallett M (1996) The role of the dorsolateral prefrontal cortex in implicit procedural learning. Exp Brain Res 107:479–485

Pavlov IP (1927) Conditioned reflexes. An investigation of the physiological activity of the cerebral cortex. Oxford University Press, London

Protopapas A, Ahissar M, Merzenich MM (1997) Auditory processing deficits in adults with a history of reading difficulties. Neurosci Abstr 23:491

Recanzone GH, Merzenich MM, Jenkins WM, Grajski KA, Dinse HA (1992a) Topographic reorganization of the hand representational zone in cortical area 3b paralleling improvements in frequency discrimination performance. J Neurophysiol 67:1031–1056

Recanzone GH, Merzenich MM, Jenkins WM (1992b) Frequency discrimination training engaging a restricted skin surface results in an emergence of a cutaneous response zone in cortical area 3a. J Neurophysiol 67:1057–1070

Recanzone GM, Merzenich MM, Schreiner CS (1992c) Changes in the distributed temporal response properties of SI cortical neurons reflect improvements in performance on a temporally-based tactile discrimination task. J Neurophysiol 67:1071–1091

Recanzone GH, Schreiner CE, Merzenich MM (1993) Plasiticity in the frequency representation of primary auditory cortex following discrimination training in adult owl monkeys. J Neurosci 13:87–103

Semel EM, Wiig EH, Secord WA (1995) Clinical evaluation of language fundamentals, Third Edition, San Antonio, TX, The Psychological Corporation

Tallal P, Piercy M (1973) Defects of non-verbal auditory perception in children with developmental aphasia. Nature 241:468–469

Tallal P, Piercy M (1974) Developmental aphasia: rate of auditory processing and selective impairment of consonant perception. Neuropsychologia 12:83–93

Tallal P, Piercy M (1975) Developmental aphasia: the perception of brief vowels and extended stop consonants. Neuropsychologia 13:69–74

Tallal P, Miller S, Fitch RH (1993) Neurobiological basis of speech: a case for the preeminence of temporal processing. Ann NY Acad Sci 682:27–47

Tallal P, Miller SL, Bedi G, Byma G, Wang X, Nagarajan SS, Schreiner C, Jenkins WM, Merzenich MM (1996) Language comprehension in language-learning impaired children improved with acoustically modified speech. Science 271:81–84

Wang X, Merzenich MM, Sameshima K, Jenkins WM (1995) Remodelling of hand representation in adult cortex determined by timing of tactile stimulation. Nature 378:71–75

Weinberger NM (1995) Dynamic regulation of receptive fields and maps in the adult sensory cortex. Ann Rev Neurosci 18:129–158

de Wierdt (1989) Spectral processing deficit in dyslexic children. Appl Psychol 9:163–174

Wright BA, Lombardino LJ, King WM, Puranik CS, Leonard CM, Merzenich MM (1997a) Deficits in auditory temporal and spectral processing in language-impaired children. Nature 387:176–178

Wright BA, Buonomano DV, Mahncke HW, Merzenich MM (1997b) Learning and generalization of auditory temporal-interval discrimination in humans. J Neurosci 17:3956–3963

Subject Index

Springer
and the
environment

At Springer we firmly believe that an international science publisher has a special obligation to the environment, and our corporate policies consistently reflect this conviction.
We also expect our business partners – paper mills, printers, packaging manufacturers, etc. – to commit themselves to using materials and production processes that do not harm the environment. The paper in this book is made from low- or no-chlorine pulp and is acid free, in conformance with international standards for paper permanency.

 Springer

Printing: Saladruck, Berlin
Binding: Buchbinderei Saladruck, Berlin